Stiegler and Technics

Critical Connections

A series of edited collections forging new connections between contemporary critical theorists and a wide range of research areas, such as critical and cultural theory, gender studies, film, literature, music, philosophy and politics.

Series Editors
Ian Buchanan, University of Wollongong
James Williams, University of Dundee

Editorial Advisory Board

Nick Hewlett
Gregg Lambert
Todd May
John Mullarkey
Paul Patton
Marc Rölli
Alison Ross
Kathrin Thiele
Frédéric Worms

Titles available in the series

Badiou and Philosophy, edited by Sean Bowden and Simon Duffy
Agamben and Colonialism, edited by Marcelo Svirsky and
  Simone Bignall
Laruelle and Non-Philosophy, edited by John Mullarkey and
  Anthony Paul Smith
Virilio and Visual Culture, edited by John Armitage and Ryan Bishop
Rancière and Film, edited by Paul Bowman
Stiegler and Technics, edited by Christina Howells and Gerald Moore

Forthcoming titles

Butler and Ethics, edited by Moya Lloyd
Nancy and the Political, edited by Sanja Dejanovic
Badiou and the Political Condition, edited by Marios Constantinou

Visit the Critical Connections website at
www.euppublishing.com/series/crcs

Stiegler and Technics

Edited by Christina Howells and Gerald Moore

EDINBURGH
University Press

© editorial matter and organisation Christina Howells and Gerald Moore, 2013
© the chapters their several authors, 2013

Edinburgh University Press Ltd
22 George Square, Edinburgh EH8 9LF

www.euppublishing.com

Typeset in 11/13 Adobe Sabon by
Servis Filmsetting Ltd, Stockport, Cheshire,
and printed and bound in Great Britain by
CPI Group (UK) Ltd, Croydon CR0 4YY

A CIP record for this book is available from the British Library

ISBN 978 0 7486 7701 6 (hardback)
ISBN 978 0 7486 7702 3 (paperback)
ISBN 978 0 7486 7703 0 (webready PDF)
ISBN 978 0 7486 7704 7 (epub)

The right of the contributors
to be identified as author of this work
has been asserted in accordance with
the Copyright, Designs and Patents Act 1988.

Contents

Acknowledgements

The editors would like to express their gratitude to the following people and institutions: Wadham College, University of Oxford, where the work was conceived and most of it was executed; Durham University, where some of the later stages of editorial work were completed; Carol Macdonald (editor) and James Williams and Ian Buchanan (series editors) at Edinburgh University Press, for their help and encouragement. We would also like to thank Austin Smidt for his unexpected invitation to participate in a highly enjoyable conference on political philosophy at the University of Dundee, where we gave early versions of our chapters and profited from the ensuing discussions. Special thanks are due to two particularly significant figures in the book's inception: Bernard Stiegler himself, whose generous sharing of his ideas with a large number of the book's contributors – in person, in writing and in his doctoral seminar series at the École de philosophie d'Épineuil-le-Fleuriel – has meant that this collection has been able to draw on the latest state of his philosophical project, including much unpublished material; and to Hugh Silverman, founder of the IPS (International Philosophical Seminar), which hosted two seminal workshops on the work of Stiegler in 2007 and 2012. Three of the chapters in this volume, those of both the editors and Serge Trottein, were first presented at the IPS at Castelrotto in the Dolomites, and benefited greatly from the discussions and criticisms they engendered there. We had very much hoped to include Hugh's essay, but were prevented from doing so by his illness and untimely death. This volume is dedicated to his memory.

We would also like to express our appreciation to the editors of *Cultural Politics* for allowing us to reprint a shortened version of an article by Richard Beardsworth, first published in 2010 (6.2, 181–199). Particular gratitude is also due to Kate and to Bernard for their tolerance of editorial meetings and lunches that may have sometimes seemed to distract us from home and its comforts. Finally, we are grateful to all our contributors not only for their excellent chapters but also for their patient cooperation with a pair of editors who at times must have resembled an uncanny hybrid of copy-editor and superego; and of course to each other for working so indefatigably – and for being such fun to work with.

CMH and GM, Oxford–Durham, June 2013

Abbreviations and Guide to Referencing

The following gives a key to Stiegler's principal texts, which have been referenced by way of abbreviations throughout.

In many cases, the abbreviations are followed by two sets of numbers, separated by a slash (/). The second set refers to the (French) original, and this is the case throughout the volume, where dual references have been given as a single bibliographical entry for both translation and original, rather than obscure the text with unnecessarily heavily parentheses. Where Stiegler references are followed by only a single set of page numbers, these refer to the translation – unless the text is only available in one language, be it English or French. Where deemed useful, they have been marked as either 'eng' or 'fr'.

In addition to those books already translated, a number of English versions of Stiegler are due to be published shortly. We have tried to anticipate their arrival by using the official translated titles in the text (for example, *Symbolic Misery* and *What Makes Life Worth Living*, both of which are forthcoming from Polity). We have also, where possible, given section numbers after the page reference, marked by [§XX] in the text, so as to enable readers better to trace citations as and when the English becomes available. Unfortunately, the English-language publishers have not always respected the section numberings of Stiegler's originals, preferring to start again from §1 at the start of a new chapter, rather than retain the continuous section numbers of the French. This can at least be solved by counting – albeit not so easily in the case of *Ce qui fait que la vie vaut la peine d'être vécue: De la pharmacologie* (*What makes Life Worth Living: On Pharmacology*), where the French inexplicably skips from §23 to §30 at the end of Chapter 3 ('On Pharmacology').

As is inevitably the case when a new thinker begins to appear in another language, there are multiple issues of translation that emerge only later, when translators' initial uncertainty gradually gives way to consensus over certain aspects of interpretation. This is exacerbated when the figure being translated is, both intellectually and stylistically, as formidably difficult as Stiegler, who tends to write by dictation, in tortuously long and syntactically complex sentences. His first wave of translators have been faced with a near-impossible task, and have unsurprisingly made mistakes. Rather than repeatedly drawing attention to these by highlighting the places in which citations have been modified, both the editors and contributors have simply retranslated where appropriate. The page numbers assigned to English translations are correct throughout – even if the words on the page are not quite the same.

AH   'Anamnesis and Hypomnesis: Plato as the first thinker of proletarianisation.' <http://www.arsindustrialis.org/anamnesis-and-hypomnesis>

AO   'How I Became a Philosopher', in *Acting Out*, trans. David Barison, Daniel Ross and Patrick Crogan (Stanford: Stanford University Press, 2008); *Passer à l'acte* (Paris: Galilée, 2003).

CE1   *Constituer l'Europe, 1: Dans un monde sans vergogne* (Paris: Galilée, 2005).

CE2   *Constituer l'Europe, 2: Le Motif européen* (Paris: Galilée, 2005).

CPE   *For a New Critique of Political Economy*, trans. George Collins and Daniel Ross (Cambridge: Polity, 2010); *Pour une nouvelle critique de l'économie politique* (Paris: Galilée, 2009).

DD1   *The Decadence of Industrial Democracies: Disbelief & Discredit, vol 1*, trans. Daniel Ross and Suzanne Arnold (Cambridge: Polity, 2011); *Mécréance et discrédit, vol I: La Décadence des democracies industrielles* (Paris: Galilée, 2004).

DD2   *Uncontrollable Societies of Disaffected Individuals: Disbelief & Discredit, vol. 2*, trans. Daniel Ross (Cambridge: Polity, 2012); *Mécréance et discrédit, 2: Les Sociétés incontrôlables d'individus désaffectés* (Paris: Galilée, 2006).

DD3   *Mécréance et discrédit, vol III: L'Esprit perdu du capitalisme* (Paris: Galilée, 2006).

DP with Marc Crépon. *De la démocratie participative: Fondements et limites* (Paris: Mille et une nuits, 2006).

DSL 'Desire and Knowledge: The Dead Seize the Living', trans. George Collins and Daniel Ross (undated). <www.arsin dustrialis.org/desire-and-knowledge-dead-seize-living>

DT 'Deconstruction and Technology: Fidelity at the Limits of Deconstruction and the Prosthesis of Faith', trans. Richard Beardsworth, in Tom Cohen, ed., *Jacques Derrida and the Humanities: A Critical Reader* (Cambridge: Cambridge University Press, 2001).

EC *États de choc: Bêtise et savoir au XXIe siècle* (Paris: Mille et une nuits, 2012).

EHP with Philippe Petit and Vincent Bontems. *Économie de l'hypermatériel et psychopouvoir: Entretiens* (Paris: Mille et une nuits, 2008).

ET with Jacques Derrida. *Echographies of Television: Filmed Interviews*, trans. Jennifer Bajorek (Cambridge: Polity Press, 2002); *Échographies de la télévision: Entretiens filmés* (Paris, Galilée, 1996).

FI with Serge Tisseron. *Faut-il interdire les écrans aux enfants? Entretiens réalisés par Thierry Steiner* (Paris: Mordicus, 2009).

L 'To Love, To Love Me, To Love Us: From September 11 to April 21', in *Acting Out* (AO); *Aimer, s'aimer, nous aimer: Du 11 septembre au 21 avril* (Paris: Galilée, 2003).

MAI 'Manifesto of Ars Industrialis', in *The Re-enchantment of the World*; 'Manifeste d'Ars Industrialis', in *Réenchanter le monde* (RW). <www.arsindustrialis.org/manifeste2005>

MAI2 'Manifesto of Ars Industrialis (2010)'; 'Manifeste d'Ars Industrialis (2010)'. <www.arsindustrialis.org/manifeste-2010>

MS 'The Magic Skin, or, The Franco-European Accident of Philosophy after Jacques Derrida', *Qui Parle*, vol. 18/1 (2009), pp. 97–110; 'La Peau de chagrin ou L'accident franco-européen de la philosophie d'après Jacques Derrida', *Rue Descartes*, vol. 52 (2006), p. 103–12.

PA with Élie During. *Philosopher par accident: Entretiens* (Paris: Galilée, 2004).

PFM with Christian Fauré, et al. *Pour en finir avec la*

mécroissance: Quelques réflexions d'Ars industrialis (Paris: Flammarion, 2009).

RW   with Marc Crépon, George Collins and Catherine Perret. *The Re-enchantment of the World: The Value of the Human Spirit versus Industrial Capitalism*, trans. Trevor Arthur (London: Continuum, 2012); *Réenchanter le monde: La Valeur esprit contre le populisme industriel* (Paris: Flammarion, 2006).

SM1   *De la misère symbolique, 1: L'Époque hyperindustrielle* (Paris: Galilée, 2005).

SM2   *De la misère symbolique, 2: La Catastrophe du sensible* (Paris: Galilée, 2005).

TC   *Taking Care of Youth and the Generations*, trans. Stephen Barker (Stanford: Stanford University Press, 2010); *Prendre soin, 1: De la jeunesse et des générations* (Paris: Flammarion, 2008).

TD   *La Télécratie contre la démocratie: Lettre ouverte aux représentants politiques* (Paris: Flammarion, 2006).

TDI   with Peter Hallward. 'Technics of Decision: An Interview', trans. Sean Gaston, in *Angelaki*, 8.2, pp. 154–68.

TT1   *Technics and Time, 1: The Fault of Epimetheus*, trans. Richard Beardsworth and George Collins (Stanford: Stanford University Press, 1998); *La Technique et le temps, 1: La Faute d'Épiméthée* (Paris: Galilée, 1994).

TT2   *Technics and Time, 2: Disorientation*, trans. Stephen Barker (Stanford: Stanford University Press, 2009); *La Technique et le temps, vol. II: La Désorientation* (Paris: Galilée, 1996).

TT3   *Technics & Time, 3: Cinematic Time and the Question of Malaise*, trans. Stephen Barker (Stanford: Stanford University Press, 2011); *La Technique et le temps, vol. III: Le Temps du cinéma et la question du mal être* (Paris: Galilée, 2003).

WML   *What Makes Life Worth Living: On Pharmacology*, trans. Daniel Ross (Cambridge: Polity, 2013); *Ce qui fait que la vie vaut la peine d'être vécue: De la pharmacologie* (Paris: Flammarion, 2010).

Introduction: Philosophy – The Repression of Technics

Christina Howells and Gerald Moore

This collection of essays is the first book in English on the work of Bernard Stiegler, an increasingly significant figure within a new generation of French philosophers that has, over the last decade or so, emerged in the wake of Foucault, Deleuze and Derrida. Although not yet as well-known as other contemporary figures such as Alain Badiou, Jean-Luc Nancy and Slavoj Žižek, Stiegler is rapidly becoming so. At the time of writing, his name is probably still most familiar from *Echographies of Television* (1996), a collaboration with his mentor, Jacques Derrida; and from the first volume of *Technics and Time* (1994), a work that reads Heideggerian technics through the neglected figures of Gilbert Simondon, a philosopher of technology and evolution, and André Leroi-Gourhan, a paleoanthropologist. His subsequent work has moved beyond philosophical anthropology into psychoanalysis, sociology and economics, as well as film, literature, art and even neuroscience. It has also made more explicit the relation of his philosophical programme to politics, expressed in his concerns over the 'catastrophic' effect of contemporary technologies on the attention span and motivation of those who consume them. This crossing of disciplinary boundaries is a hallmark of Stiegler's work, and has led him to move beyond academia into the public arena. Since the turn of the century, he has grown into the (Sartro-Foucauldian) role of the public intellectual, both 'committed' and 'specific', participating in high-profile debates on matters as diverse as terrorism and 9/11, the rise, in France, of the National Front, and the French presidential elections of 2002 and 2007 (L, TD), the causes of the 2008 financial crisis (CPE), the effect of advertising and screen technologies on the development of young children (FI), and the kinds of massacres carried out by Richard Durn in Nantes, Anders Breivik in Utøya and Mohamed Merah in Toulouse.[1]

Stiegler's response to these issues can best be summarised by the campaign line of the think tank he founded in 2005, Ars Industrialis, which states the need 'for an industrial policy on technologies of the mind' (RW), or, in other words, a politicisation of the question of the relationship between technology and mental, or spiritual, vitality. Stiegler argues that Western democracies are plagued by a 'a crisis of the spirit' and 'symbolic misery', consisting in apathy, disaffection and social collapse, and that this crisis is a product of mass media and the technologies of consumerism to which we have become addicted, and on which we depend for the construction of both society and ourselves. But he also insists that it is also only through technology that social *re*construction – over and above Derrida's *de*construction – can take place; that we can hope to 're-enchant the world', as he puts it in the book of the same name. The breadth and contemporaneity of these concerns helps account for Stiegler's emergence as a crucial theoretical point of reference for an understanding of the intellectual and political landscape of the twenty-first century.

The title of this volume, *Stiegler and Technics*, obviously implies a focus on Stiegler's works on technology, but it should already be apparent that this cannot be kept distinct from his engagement with other disciplines, nor from the critique of consumer capitalism that stands as Stiegler's other major contribution to contemporary philosophy and social theory. Technics is at the heart of what it is to be human for Stiegler, and this means that any change to the technical environment that constitutes culture is of potentially huge significance for the nature of human existence – all the more so when that change takes place at the rate we are currently witnessing. On a philosophical level, technics provides a less narrow focus than might at first seem likely. Contemporary philosophers of technology tend to begin with Martin Heidegger's seminal 'Question Concerning Technology' (1954), and Stiegler is no exception, though he challenges Heidegger's central claim. Heidegger famously described 'the essence of modern technology' as being 'itself nothing technological' (Heidegger 1993: 324–5), meaning that an understanding of what technology *is* cannot be deduced from an analysis of technology itself, that is, from gadgets and other technical artefacts. For Heidegger, technology is significant not because of what it tells us about technology, but because of what it tells us about *ourselves*, our 'ontological' way of being in the world. In the modern era, he argues, this relationship with

Being has been forgotten; our relation to the world has been predominantly negative, or nihilistic. We have employed machines to extend the dominion of man over nature, rather than reconcile ourselves to the inevitability of our death. And we have used technics to impose a restrictive 'order' (*logos*) on life, 'enframing' the earth in a technical-industrial apparatus that refuses it the possibility of withdrawing from our grasp (1993: 332–3). Technology has thus become an instrument for the immortalisation of humanity – and one that, in the process, has set about destroying the planet on which our survival depends. This was not always the case, however. Heidegger contrasts the ontological impoverishment of modern technology with a long-forgotten, pre-technological, Ancient Greek concept of *tekhnē*, meaning art or poetic craft, in which he finds the fullest affirmation of our mortality (1993: 318, 339). Embodied in poetic depictions of nature's refusal to submit to the ordering imposed on it by human reason, this classical conception of technics as the 'unveiling', rather than the framing, of Being, served precisely as a recognition of the very human finitude that, according to Heidegger, modern technology attempts to deny.

In a key section of his first book, *Technics and Time, 1: The Fault of Epimetheus* (1994), Stiegler builds upon this Heideggerian insight, returning to Ancient Greek mythology to find a way of thinking about technics that has not yet succumbed to corruption by the modern metaphysical obsession with humanity's privileged place in the world. Michael Lewis explores some of the key philosophical reasons for the recourse to myth in one of this volume's earlier chapters. But Stiegler also uses mythology to argue that Heidegger's positive concept of ancient *tekhnē* still falls short: Heidegger envisages a kind of poeticised technics that gives more truthful expression to the kind of finite beings that we are, but he does not see that we *are* – that we exist as humans – *only through technics*. In other words, Stiegler calls into question Heidegger's claim about the truth of technics being found outside technology, in the consciousness of being-in-the-world that Heidegger terms Dasein (TT1, 18/31). Human consciousness, Stiegler argues, is always already technical, made possible by technics. This is because it is only through technics that we create time, inventing ourselves a future through the inheritance of acquired experience and the horizons of expectation to which this gives rise. We are defined and moreover *constituted* by an externalised memory of a past that we never lived, namely culture, which is composed of technical objects that embody the knowledge of our ancestors, tools that we adopt to transform our environment, enabling us to anticipate and stave off death.

Drawing on the work of the classicist Jean-Pierre Vernant, Stiegler locates the originary sense of technics in Hesiod's *Theogony* and *Works and Days*, two poems dated to around 700 BCE, which set out the origins of the cosmos and of humans respectively. Hesiod is the earliest source of the myth of Prometheus and his routinely overlooked twin, Epimetheus, the Titans credited with having created the first humans. The poet relates how, soon after the creation of the world, Zeus charged the twins with handing out gifts, or qualities, to all the creatures, so that they might co-exist without inter-species domination. Having shared out qualities like fur, for keeping warm, fleetness of foot, for escaping predators, and sharp eyesight, for spotting both enemies and prey, Epimetheus discovers only too late, when all the gifts have already been distributed, that he has forgotten about man, who is left with nothing. To borrow a phrase from Robert Musil, man is accordingly 'without qualities' (TT1, 193/201), defined less by what he has than by what he lacks – since what he intrinsically has is precisely *nothing*. Prometheus tries to make up for his brother's oversight by stealing fire from Mount Olympus to give to humans, for which he is famously punished by being chained to a rock while his liver is devoured daily by an eagle. In spite of this, the gift of fire is what enables humans to survive. In Stiegler's terms, it gives rise to the 'invention of the human'; but in this respect it also portends our susceptibility to downfall. Crucially, it is not a quality, since it does not form part of an essence that determines what we are. It is rather a supplement that stands in for the essence that we lack.

Fire is the first technics, an artificial, technical prosthesis, or externalised organ, that enables but also condemns man to live outside of himself. An instance of what Stiegler will later term the *pharmakon*, it is both cure and poison, our salvation and our ruin: what lifts humans above animals, but also a symbol of mortality, which sees us banished from dining at the table of the gods (TT1, 194/201–2; DD3, 115). The poisonous dimension of technics is made clear from the outset, when, coinciding with the punishment of Prometheus, Zeus also punishes man for being a receiver of stolen goods. Another poisonous gift is sent in the form of Pandora, the first woman, 'a beautiful evil' (Vernant 1989: 62/116), whose arousal of fire in man's heart leaves him henceforth tormented by unquenchable desire. For Stiegler, this continuation of the story reveals desire to be the inner correlate of our externalisation in technics, the affect emerging in response to our technical organisation. It thus exemplifies what is perhaps the single most decisive tenet of his philosophical thought, namely that we are nothing but the internalisation of our prosthetic *ek-sistence* (TT1, 152/162; WML, 41fr [§9]). Selfhood, or interiority, emerges at the site of our nothingness, with the life of the mind (*esprit*) created, or, as he will say, *individuated*, by the technical objects through which our existence is externalised: '*the who is nothing without the what*' (TT2, 6/14). This concept of individuation is taken from the evolutionary philosophy of Gilbert Simondon, for whom it describes the way that an individual and its environment (or 'milieu') do not pre-exist one another, but are rather entwined in an ongoing, ontogenetic relationship of mutual constitution, or 'transduction'. According to Simondon, 'mental function [*le psychisme*] is neither pure interiority nor pure exteriority, but permanent differentiation and integration' of the social and technical milieu in which the mind is suspended (Simondon 2007: 98, see also 12–17). The implication is that the human brain is no more than a surface of inscription on which mind, or spirit, is written through the tools we use to navigate the world, though as Ian James points out, in a chapter that puts Stiegler into dialogue with the work of Catherine Malabou, a developed theory of cerebrality is currently missing from Stiegler. The lacuna looks set to be filled however, by the long-awaited, as yet unpublished, *Technics and Time, 4*.

In the twenty years since the first volume of *Technics and Time*, Stiegler has written some thirty books, which take the initial thesis of the technical constitution of the human and apply it increasingly to the spheres of politics, economics, aesthetics and science. A rough distinction might be made between the first, early phase of his work, on the concept of technics; a second phase on what he calls 'libidinal economy'; and a more recent third, in which he reworks both technics and libidinal economy in terms of the concept of 'pharmacology'. The second period of output, from around the time of the first volume of *Disbelief and Discredit* (2004), is particularly concerned with elaborating the relationship between technics and desire, to which we shall return. His first and opening phase, comprising the first three volumes of *Technics and*

Time (1994–2001), is more specifically concerned with the idea that humanity is 'invented' through technics – and that philosophy, since its inception, and up to and including Heidegger and Derrida, has failed sufficiently to understand this.

Twentieth-century philosophy often seems to start with the claim of having discovered a decisive concept that was forgotten or repressed with the advent of metaphysics. As is well known, Heidegger's *Being and Time* (1926) opens with the intention to 'restate the Question of Being' (Heidegger 1962: 21 ¶1), which he suggests had been forgotten since the dawn of a philosophical modernity that, like Nietzsche, he identifies with Plato. Gilles Deleuze similarly described his philosophical project as a 'reversal of Platonism', uncovering a positive conception of difference, which the metaphysical tradition had hitherto construed purely in terms of privation and lack. At much the same time as Deleuze, Jacques Derrida announced the arrival of deconstruction with the claim, once again, that philosophy since Plato had been predicated on an incorrect assumption of the privilege of (living) speech over (dead) writing.[2] For Stiegler, it is not Being or difference or writing that has not been thought, but technics. As he states in the preface to *The Fault of Epimetheus*: 'at its very origin and up until now, philosophy has *repressed* technics as an object of thought. Technics is the *unthought*' (TT1, ix/11). A later interview formulates the claim even more boldly:

> The question of philosophy is *entirely*, and since its origin, that of the endurance of a condition that I call techno-logical: simultaneously technical and logical, initially forged on the axis that forms language and tools, which is to say, on the axis that enables man's *externalisation*. [. . .] Since its origin, philosophy has been marked by this techno-logical condition, *but via its repression* [refoulement] *and denegation*, and the difficult project I have undertaken is to show that *philosophy begins with the repression of its ownmost question*. (PA, 14–15)

Throughout its history, philosophy has been a discourse on the one question whose import it has declined explicitly to acknowledge, namely technics. It begins with the separation of *logos*, denoting a privileged concept of reason, or logic, from *tekhnē*, which is linked to (implicitly inferior) practical skills. Their separation coincides with the consolidation and re-entrenchment of writing over the course of fifth-century Athens. Prior to this, in the broadly preliterate, oral and sacrificial culture of Homer and Hesiod, *tekhnē* and *logos* were one and the same. Knowledge was inherently practical, embodied in the myths and rites of sacrifice, through which men acknowledged their mortality and their inability to compete with the gods. The abstract nature of written language brought with it the discovery of universal laws that appeared to transcend experience. A development in the history of technics, in other words, served as the condition for the advent of philosophy. Plato himself refused to see this, criticising writing as an instrument for the corruption of meaning and the weakening of memory, a mere 'appearance of wisdom' in which the process of externalisation brought about the loss of inner truth (*Phaedrus*, 274e–275e). But, Stiegler argues, the abstract thought processes that writing makes possible lie behind Plato's invention of the Ideas (justice, beauty, and so on), a realm of metaphysical truth that exists independently of the world. The ability of men to access this realm through thought elevated us to the level of gods, able to 'know' the once mysterious mind of a now logical cosmos.

Like Nietzsche, Heidegger, Derrida and Deleuze before him, Stiegler identifies the simultaneous beginning and repression of philosophy with Plato. He accords a particular place to the *Meno*, a dialogue in which, according to Stiegler, we see Plato emerge from the shadows of Socrates, his teacher, who belongs, as its last stand, to a (Hesiodic) age of tragedy that 'is *not yet* philosophy' (PA, 25–6). The decisive episode in the *Meno* is the exposition of the claim by Socrates that true knowledge is not – indeed, cannot be – acquired through teaching and experience, but rather consists in the 'recollection', or *anamnesis*, of eternal, *a priori* truths. Socrates famously brings Meno to agree with him by way of a series of geometrical puzzles, to which the young slave boy works out the answer despite having no previous experience of geometry. Stiegler's interest in the dialogue lies in precisely what Plato leaves out from his explanation. Although the text makes no reference to writing or drawing, it is apparent from the context that Socrates is tracing out the diagram of a square by way of demonstration, externalising his thought through technical prostheses, which serve in turn as the condition of Meno's intellectual awakening (82b–85e). The diagrams function as what Stiegler calls '*hypomnemata*', external memory supports, without which there can be no interiority, or 'recollection' of knowledge: '*anamnesis* (reminiscence) is always supported and inhabited by a *hypomnesis* (a mnemotechnics)' (PA, 38). Or, to put it in Kantian terms, the externalisation of memory in technics is the ('transcendental') condition of the possibility of experience (PA, 32). Clouded by the desire, articulated in the *Phaedrus*, to see writing as an evacuation, or displacement, of thought into our hypomnesic prostheses, Plato is unwilling to acknowledge that the process of externalisation is actually what makes thought possible. By insisting that philosophical knowledge is 'recollected' from memory, without prompting from outside, he divorces *ēpistēmē* and *logos*, knowledge and reason, from *tekhnē*, the external memory supports, the cultural artefacts through which we are able to think in the first place (TT1, 1/15).

There are rare exceptions, such as Foucault and Simondon, whose work on *hypomnemata* and technical evolution, respectively, are discussed by Sophie Fuggle and Stephen Barker in this volume. For the most part, however, Stiegler argues that Plato's denegation and repression of technics is repeated right across the history of philosophy, which accordingly plays out as a 'forgetting of Epimetheus' – a forgetting of humans' originary condition of default; of the way that we come to exist as humans, to think, only through technics. Kant's mistake, according to Stiegler, is to have located the transcendental conditions of experience, the categorical rules through which we convert raw sense data into the identifiable objects and object-relations of experience, in hardwired, *a priori* structures of subjectivity, rather than outside us in the technical organs of the externalised human mind. As Patrick Crogan shows, explaining Stiegler's 'cinematic' understanding of time, this critique also extends to the phenomenologist Edmund Husserl, who, according to Stiegler, overlooks the role played by external 'tertiary memory' in the construction of anticipation. Daniel Ross's essay takes us beyond Husserl to Derrida, whom Stiegler similarly criticises for failing to recognise that meaning can be transformed in the future only if we preserve the material traces of its history – without which there will be no future at all (DT, 238–44). Stiegler also takes the psychoanalyst Sigmund Freud to task for locating the origins of the unconscious in biology, rather than in the technical memory of culture, which creates desire by placing obstacles in the way of our satisfaction. As Tania Espinoza explains in her chapter, Stiegler draws on the work of Donald W. Winnicott to argue that we learn to desire through technics. This, in turn, provides the basis for his argument that economics and politics are inherently libidinal, that is to say, a matter of the organisation of desire through technics.

A number of other chapters in this collection are more critical of Stiegler's readings of the history of philosophy, raising issues with what Oliver Davis, writing on Stiegler's use of the Freudian concept of sublimation, describes as his 'somewhat predictable' criticism of other thinkers' blindness to the effects of technology (p. 171 of this volume). Approaching Stiegler via the question of the space that philosophy-as-technics leaves open for aesthetics, Serge Trottein asks whether externalised memory really performs the function that Stiegler criticises Kant for neglecting. Christopher Johnson returns to André Leroi-Gourhan, whose work on the relation between human evolution and the external memory of technology provides the basis for Stiegler's own account, but whom Stiegler suggests did not himself quite grasp the technical constitution of the human. In what emerges as a common theme regarding the attentiveness, or lack thereof, with which Stiegler reads his various interlocutors, Johnson suggests that this conclusion is a product of Stiegler's own rather compressed reading. Christina Howells makes a similar point about Stiegler's use of Lacan, in a chapter on the erosion, or 'desublimation', of desire by consumer capitalism.

In contrast to the patient, diligent deep-readings we have come to expect since Derrida, the image that emerges of Stiegler is perhaps of a thinker who zooms in and out of texts, skimming them to fit what he is looking to find – as if afflicted by the very 'hyper-attention' he laments as becoming endemic. The montage-style of his tortuously long sentences, which frequently 'cut' from one part of the history of philosophy to another, and between contemporary politics, science and technology, doubtless adds to the sense of reading as if clicking through hyperlinks, laid out by an eclectic, impatient writer. But these criticisms of some of the details of Stiegler's project need not derail it. On the contrary, they show Stiegler to be a product of the society whose excesses he is challenging. Viewed from another angle, this has been one of his greatest strengths. Famously, Stiegler came to philosophy late, at the age of twenty-six, while serving a stint in prison for armed robbery (1978–1983). In the autobiographical *Acting Out* (2003), he describes how being deprived of his habitual environment, of the external memory supports through which he previously lived, meant that he also lost his sense of self, and had to begin again as if from scratch. He rebuilt his interiority through reading, with the (philosophy) books that he read serving as an external memory he would internalise as the basis of his thinking. It was this, he suggests, that led to his discovery of the technical constitution of the human. And it is also this that perhaps explains why, unlike the generation of French philosophers who preceded him, including Deleuze, Foucault and Jean-François Lyotard, Stiegler is reluctant to see the institutions of society as unilaterally oppressive regimes for the containment of desire. Poststructuralism has tended to romanticise the loss of self, valourising the breakdown of subjectivity as a way of opening experience onto hitherto unimaginable intensities of affect; advocating the 'deterritorialisation' of social structures as a way of bringing us into contact with an 'event' of life that is routinely policed out of existence, kept at bay on account of its revolutionary potential. Deleuze and Guattari's *Anti-Oedipus: Capitalism and Schizophrenia, 1* (1971) and Lyotard's *Libidinal Economy* (1974) are arguably the foremost examples of this, theorising capitalism and the institutions of political economy as manipulations of otherwise subversive desire, which is tamed by being channelled into docile subservience and profit (Deleuze and Guattari 2004a; Lyotard 2004). Stiegler agrees with their assessment of consumer capitalism as the production of lack, the exploitation of desire, but rejects the idea that desire could exist – let alone exist untrammelled – in the absence of social structures. For Stiegler, the organisation of desire is precisely what makes it possible. The collapse of selfhood is not a fantasy to be encountered beyond the limits of capitalism, but a reality found at its centre, borne out in spiralling rates of apathy, disaffection, and an increasingly prevalent 'feeling of not existing', of the kind described by Richard Durn (DD2, 83–90/121–30).

In a prolific period of output comprising the two volumes of *Symbolic Misery* (*De la misère symbolique*, 2004) and the three-volume *Disbelief and Discredit* (*Mécréance et Discrédit*, 2004–6), as well as the two-volume *Constituer L'Europe* (*Constituting Europe*, 2005) and numerous smaller works, Stiegler returns to and develops an earlier, Freudian account of 'economics of the libido', meaning the ways in which libidinal energies are invested and withdrawn from particular objects (Freud 1961: 80n; Freud 1957: 74–5). *The Decadence of Industrial Democracies* opens with the (typically italicised) claim that *'libidinal economy is the deep-seated mechanism of all adoption'* (DD1, 10/27). This means that desire determines which technics are adopted by a society and its individuals – and moreover that the nature of the technics adopted will determine, via the libidinal investments undertaken, the very forms of our desire. Reworking Derrida's concept of *différance* as a theory of the 'deferral of satisfaction', Stiegler argues that desire is created by institutions – externalised bodies of technical knowledge – that defer the prospect of pleasure, and in the process create anticipation, through the projection of long-term horizons of expectation (DD2, 56–7/84–5). The problem with consumerism is that it does the very opposite of this. Marketing technologies' relentless promotion of consumption as a means to immediate gratification has 'short-circuited' the processes of deferral through which desire is sublimated. Seemingly unlimited access to pleasure has led to the annihilation of long-term goals, leaving us addicted to the short-term quick fix, unable to envisage a future beyond the immediate present (CPE, 57–9/81–3).

This aspect of Stiegler's work has drawn criticism from a number of contributors to this volume. In his chapter on *For a New Critique of Political Economy* (2009), published in the aftermath of the 2008 credit crunch, Miguel de Beistegui, for example, wonders whether Stiegler's position does not ultimately fall back on a metaphysical Freudian model of desire that is itself a product of capitalism. Richard Beardsworth suggests that the linking of desire to technology lends itself to suspicions of technological determinism, though Ben Roberts pointedly rejects this suspicion in the chapter that follows, on what Stiegler calls the 'lost spirit of capitalism'. The phrase, recalling Max Weber's disciplined, vocational entrepreneur (DD3, 18; Weber 1958), hints at others' concerns over Stiegler's politics, notably his apparent nostalgia for a bygone age of social order and pre-consumer capitalism. With his emphasis on the conservation of tradition and society's need for a sublimating superego, Stiegler risks seeming disconcertingly reactionary for a supposedly radical poststructuralist. Indeed, despite his critique of Plato's attitude toward technics he frequently repeats the *Phaedrus*'s line about writing as the 'production of forgetfulness' (274a), reworking it into a critique of the anaesthetising gadgetry of the twenty-first century. Long before Marx, Plato becomes the *'first thinker of the proletariat'*, theorising the debasement and dehumanisation of the human spirit (CPE, 28/43); his concerns over the dangers of displacing live thought into the dead matter of technics (writing) anticipate the liquidation of traditional skills of craftsmanship by workplace mechanisation, not to mention the phenomena of calculators making us innumerate, mobile phones emptying our brains of phone numbers, and global-positioning systems eroding our memory for directions.

Stiegler's critique of technology is only one side of the story, however. The other half pertains to his positive conception of technics, and his related efforts to create new forms of spirit (and not merely revitalise an old one). Stiegler's problem with Plato is not that he simply rejects the philosophical significance of technics, but that his thinking of technics is not pharmacological, and does not recognise – as Derrida would later put it – that the conditions of possibility and impossibility are one and the same; that the *hypomnemata* that incline us to forget are also precisely what enable us to remember, hence a cure for the toxic malaise into which they have led us. Stiegler's argument that the mind is nothing but the internalisation of our technical prostheses complicates the idea of thought being evacuated into technology. Surely the machines that nowadays think on our behalf do not so much destroy thought as create new ways of thinking, freeing up the mind to work at unprecedented levels of complexity? His answer to this question is that the inventiveness of technologies, the extent of their ability to individuate the mind, depends on our participation in them, on our ability to operate and understand them. Where Marx saw proletarianisation in terms of the inability to consume (enjoy the profits of) what we produce through labour, Stiegler locates it in 'the capture of attentional flows by the tertiary retentions that constitute *pharmaka*', and the displacement of cognitive functioning into machines (WML, 90fr [§29]). Under consumerism, this amounts to our inability to produce what we consume. Surrendering our brains for reprogramming by advertising, we unquestioningly accept whatever the market thrusts in our direction, and have, for the most part, become passive and superficial consumers rather than active producers of technology. In a cultural climate that privileges consumption over engagement and participation, we are routinely locked out from seeing the workings of our machines under the guise of 'user-friendliness', cynically promoted by manufacturers keen to protect their commercial interests. We find ourselves using computers with no idea of how they work, reading webpages without any curiosity as to how we might construct them ourselves. We are increasingly unable to use technology to invent futures that differ from those envisaged for us by the market. This is not just an issue of memory and forgetting, but of humanisation and dehumanisation. As Gerald Moore shows in this volume's opening chapter, technics, for Stiegler, marks the difference between man and animal, accounting for a break in evolution that dates back and gives rise to the earliest humans. Unlike animals, which simply live or die, depending on their evolutionary fitness, humans use technology to create artificial organs that overcome maladaptation. If contemporary capitalism is dehumanising, Stiegler argues, it is because it destroys the traditions, the accumulated experience, in which these artificial organs consist; because it liquidates the technical-cultural memory that, by enabling us to think and anticipate, gives rise to our humanity.

Stiegler's solution to what he perceives as an unprecedented crisis of humanity will be an 'economy of contribution', composed not of brands and marketing, but of what Martin Crowley's chapter describes as 'amateurs', individuals who use technology not to 'tune in and drop off', but to participate actively in the reconstruction of symbolic order. The writings of Stiegler-the-philosopher are increasingly matched by his contribution as a citizen-amateur, in this respect. Through the lobbying of Ars Industrialis and, more recently, the open-access school of philosophy he has founded in the commune of Épineuil-le-Fleuriel,[3] not to mention appearances on television and before the Commission of the European Union, he is proactive in using philosophy to re-engage with and bring together an increasingly atomised political community. The ethos of the amateur, meaning 'lover', as he sees it, is inextricable from the *philia* of philosophy – the love of knowledge that is also a love of technical *savoir-faire* and *savoir-vivre*, the know-how and life skills that we inherit, and pass on through our participation in society, and which also thereby make us human.

Notes

1. The perpetrator of the so-called 'Nantes massacre' in March 2002, Durn has been a constant reference for Stiegler, on account of a suicide note attributing his actions to a desire to acquire a 'feeling of existing' (see L, 39/13; WML, 16fr [Introduction]; see also Moore 2013). On Breivik and Merah, see Stiegler's article in *Le Monde* (20th March 2013), 'Ces abominables tueries peuvent s'expliquer par la dérive de nos sociétés' ('These abominable killings can be explained by the drift of our societies'), also available online: <www.lemonde.fr/idees/

article/2012/03/29/ces-abominables-tueries-peuvent-s-expliquer-par-la-derive-de-nos-societes_1677650_3232.html> (last accessed 16 February 2013)

2. See, for example, Deleuze 2004a: 71/82–3; Deleuze 2004b: 291/292; Derrida 1997: 17, 27/29, 42.

3. See <http://www.arsindustrialis.org> and <http://pharmakon.fr/word press/>.

I: Anthropology – The Invention of the Human

Adapt and Smile or Die!
Stiegler Among the Darwinists

Gerald Moore

The global ridicule in which the works of Foucault, Lacan, Derrida and Deleuze had suddenly foundered, after decades of inane reverence, far from leaving the field clear for new ideas, simply heaped contempt on all those intellectuals active in the 'human sciences'. The rise to dominance of scientists in all fields of thought became inevitable. [. . .] They truly believed only in science; science, to them, was a criterion of unique, irrefutable truth. [. . .] They believed in their hearts that the solution to every problem – whether psychological, sociological or more broadly human – could only be technical in nature. [. . .] MUTATION WILL NOT BE MENTAL, BUT GENETIC.

Michel Houellebecq, *Atomised*[1]

The dénouement of Michel Houellebecq's *Atomised* (1998) depicts a world in which biotechnological breakthrough has put an end to philosophy and the so-called 'human' sciences. The narrator of the epilogue – a charismatic spokesman for the genetic revolution – tells us that, by 2013, science will live on as the only remaining source of truth, able to engineer reality in a way that obviates the need for philosophical acceptance of, or reconciliation with, the iniquities of existence. A later novel, *The Possibility of an Island* (2005), goes further still, suggesting that contemporary genetic science has the capacity to release us from 'the Darwinian fiction of "the struggle for life"' (Houellebecq 2006: 282/320). In a world where the human condition can be biotechnologically rewritten, philosophy's attempts to keep up look parodic.

A firm response to this claim is found in the work of Bernard Stiegler, for whom technics does not constitute the exhaustion of philosophy, but rather its principal concern. It is nonetheless only with the biotech revolution, Stiegler argues, that this concern has become fully apparent; that philosophy, understood as a discourse on the conditions of human interiority (PA, 14–15), can finally reveal what has always already been the case, namely that Darwinian ' "laws of evolution" have been suspended since at least the invention of the human, which is to say, since technics' (TT2, 152/176–7). Ironically, though, if Darwinism is a 'fiction' of sorts, we have discovered as much only now, as the theory of evolution by natural selection threatens to become the metanarrative of a post-philosophical age of technoscience. Perhaps surprisingly, Stiegler suggests that recent French thought has bordered on complicity in this narrative, on account of an ethos of resistance that lends itself to an affirmation of adaptation (Houellebecq's 'mental mutation'), as if there were no alternative but to reconcile ourselves to the 'nature' of capitalism. 'If we have still to inherit the thoughts of those thinkers who have disappeared', we can do so, he writes, only *by breaking with the discourse of "resistance" through the question of invention* (EC, 134 [§30]).

From his earliest work in the first volume of *Technics and Time* (1994), to the most recent *États de choc: Bêtise et savoir au XXIe siècle* (*States of Shock: Stupidity and Knowledge in the 21st Century*, 2012), Stiegler develops the argument that Darwinist adaptation by natural selection is not so much a fiction as an incomplete account of the evolution, or what he calls the *invention*, of the human. He draws on but also challenges the Nobel Prize-winning molecular biology of François Jacob, whose *The Logic of Life* (1970) defines genetic evolution in terms of its impermeability to experience. According to Jacob, when genes are transmitted, they simply live or die depending on their extent of their adaptation or 'fit' with the environment. Stiegler posits that human life breaks with this, because it is defined by its response to experience, and by our use of technology to create artificial environments that overturn maladaptation. While man evolves genetically in so far as he is an animal, only through technical evolution, or 'epiphylogenesis', does he become human (TT1, 154–5/163–4). The selection process of technical evolution is one of *adoption* rather than adaptation, with the adoption of technics giving rise to what Stiegler calls 'la vie de l'esprit', the life of the mind or spirit (MAI, Art. 1). Our failure to see this distinction has been exacerbated by the adaptationist ideology of contemporary capitalism, which suppresses adoption – the possibility of inventing an alternative future – and naturalises adaptation as the defining characteristic of human society. The kinds of technology promoted by industrialisation, in both productionist and consumerist phases, mean that 'the process of adoption has been short-circuited and replaced by a process of adaptation' (WML, 210fr [§71]). The result of this 'entropic', 'economic and social Darwinism' is the short-circuiting of human life, the reduction of humans to proletarianised 'bêtes', stupid beasts, programmed to aspire to no more than hollowed-out, meaningless survival (CE2, 64 [§8]).

Critics have long argued that Darwinism needs to be understood as the product of a philosophical tradition of individualism stretching back as far as Hobbes, and which has routinely elided the role of culture in human evolution. The classical statement of this position is found in the claim of the anthropologist Marshall Sahlins that 'the theory of sociobiology has an intrinsic ideological dimension, in fact a profound historical relation to Western competitive capitalism' (Sahlins 1976: xii). Biologists, too, have expressed a profound disquiet over the economic thinking that underpins the 'ideological' metaphysics of 'universal ultra-Darwinism' (Rose 2005: 210–12, 278; see also de Waal 2009: 3–4, 38–9). To some extent, Stiegler is thus just the latest in a long line to criticise Darwinism from the standpoint of humanism. But he differs from his predecessors both by being more nuanced in his humanism and by arguing that the application of Darwinism to human society goes beyond a particular strand of liberal-conservative thought. According to Stiegler, if we do not recognise the invention of humanity through technics, it is because the whole history of Western metaphysics has been marked by the *'the repression* [refoulement] *and denegation'* of our technological condition (PA, 15) – the attempt to conceal from ourselves the extent of our prosthetic fragility *qua* humans. The popularised discourse of what Sahlins calls 'folk' biology lends itself to this repression by implying the hardwiring of identity in DNA, not to mention the ready conflation of desire with the biological urge toward sexual reproduction. For all its rejection of biological reductionism, post-structuralist philosophy is similarly guilty of repression. Deleuze, Foucault et al. pride themselves on a Nietzschean affirmation of life as an event irreducible to biology, but in so doing they shift the emphasis of living away from the invention of humanity, and onto the joyous affirmation of resistance – or subjective adaptation – to a hostile environment. The challenge they pose of 'becoming other' (*devenir-animal*, or -*Übermensch*) risks being indistinguishable from a far more problematic *devenir-bête*, heroising what amounts to survival in a willing performance of economic Darwinism. Boiled down to Nietzschean clichés of 'loving the poison' and 'that which does not kill me makes me stronger', the big names of French philosophy 'rationalise and ultimately legitimate' a regressive collapse into dehumanising bestialisation (EC, 132–3 [§30]). For Stiegler, by contrast, humanity is not a condition to be transgressed and overcome, but rather to be recognised in its prosthetic fragility, and taken care of.

Life and 'life by other means'

In the twentieth century's most prominent work on the subject, *What Is Life?* (1944), the Nobel physicist Erwin Schrödinger coined the term 'negative entropy' or 'negentropy' to describe the organisation of matter through which life avoids 'the rapid decay into the inert state of "equilibrium"', the dissipation of energy that the Second Law of Thermodynamics calls 'entropy':

> A living organism continually increases its entropy [. . .] and thus tends to approach the dangerous state of maximum entropy, which is death. It can only keep aloof from it, i.e., alive, by continually drawing from its environment negative entropy [. . .] (1992: 70–1)

Life, Schrödinger argued, is the maintenance of order against a universal tendency toward ebbing into disorder. The organism 'feeds upon negative entropy', bringing in highly organised energy through breathing, eating, and so on, so as to compensate for the heat emitted by its gradual entropic breakdown.

In order for organisms to preserve their basic structural integrity across generations, Schrödinger also posited that there must be a basis for transmitting negentropy, a means of inheriting organisation that could withstand decay and dissipation, on account of being 'largely withdrawn from the disorder of heat motion' (Schrödinger 1992: 85). His concept of 'aperiodic crystals' served as a forerunner of the DNA macromolecule containing the programme code for metabolic exchange, but not itself interacting with its environment. In the words of a later Nobel laureate, the molecular biologist François Jacob, the genetic 'programme' is impermeable to the influence of external factors: 'The programme does not learn from experience'; 'the nucleic acid structure [. . .] is inaccessible to experience and remains unchanged throughout generations' (1974: 3, 9/11, 17). Although genes are susceptible to random mutation, or mistakes in the process of replication, mutation is not an active response to environmental change. According to the received wisdom of neo-Darwinism, adaptation to circumstance does not proceed through the (Lamarckian) acquisition of subsequently heritable characteristics, but is rather a matter of random variation and passive 'fluke' (Dawkins 2006: 242; Rose 2005: 217). Adaptation by natural selection means only accidental survival, followed by the chance ability to reproduce in a particular environment.

Bernard Stiegler contests the sufficiency of neo-Darwinism for an account of the evolution of the human. His project, broadly defined in the general introduction to the *Technics and Time* series, is to supplement explanations of biological adaptation with an account of the *technical* evolution of what he later comes to call 'la vie de l'esprit' (MAI, Art. 1). 'As a "process of externalisation", technics is the pursuit of life by means other than life' (TT1, 17/31). It takes life beyond the narrowly biological and into the more amorphous sphere of culture, 'by inscribing negentropy *outside* of the living being, and as a characteristic of its vital milieu' (L, 85 n.10/20). Whatever can be said of man as a *physical* organism, Stiegler's crucial contention is that *esprit* evolves technically, rather than biologically – and that this entails a quite different mechanism of selection: not just accidental variation (mutation), chance survival and retroactive adaptation, but 'imaginary variation' and an anticipation of change that enables us to break with survival of the fittest (TT2, 150/175). Understood as the tools through which an organism is able to modify its surroundings, technics mean that adaptation no longer takes place simply at the level of the gene. Stiegler thus denies that the concept is adequate for an understanding of human *esprit*, suggesting that it is even hostile to the vitality of the psyche: 'I contend, on the contrary, that adaptation is a cause of death and entropic' (CE2, 59 [§7]).

It would be easy but quite wrong to read into this a spiritualist nostalgia for mind–body dualism. Stiegler is a materialist focusing on the evolution of human *esprit*, albeit via a different kind of evolution. Returning to the basic formulation of technical evolution in a later chapter of *The Fault of Epimetheus*, 'Who? What? The Invention of the Human', he borrows from the work of the paleoanthropologist André Leroi-Gourhan to elaborate on the idea that human life is characterised by the externalisation of memory in technics. Writing in the 1960s, Leroi-Gourhan sought, through a qualified reworking of Lamarck, to counter the popularised Darwinist commonplace of man descending directly from the primate, as if no more than the outcome of a gradual process of genetic refinement (Leroi-Gourhan 1993: 8/17). 'The human is not a spiritual miracle that would suddenly belong to an already given body, in which the "mental" is grafted onto the "animal". Man does not descend from the monkey', as Stiegler paraphrases (TT1, 144/154). Rather, he emerges from an evolutionary break that Leroi-Gourhan identifies with the australopithecine *Zinjanthropus* (*Paranthropus boisei*, 2.6–1.2 million years ago). The decisive break occurs not with the brain, but with the foot, whose flattening, coinciding with upright walking, freed hominids' hands for tool use and liberated for speech the face that had once predominantly been used for grasping at food (1993: 19–20/33–4). The ensuing history of 'hominisation' is a direct result of the use of tools to manage our environment. The corticalisation of the human brain is made possible by the development of technics, or non-corporeal organs for living, which mark the hominids' departure from purely biological evolution.

Technical evolution involves the invention of new, externalised organs that outlive their creators, passing on to subsequent generations as a kind of social-technical memory through which the acquired characteristics of society are made heritable. Writing in opposition to Lamarck, the nineteenth-century biologist August Weismann differentiated between two kinds of memory: one consisting in the genetic stock of a species, and the other being the (conventional, psychological) memory of the individual, whose accumulated experience is condemned to die alongside its bearer without ever passing into the gene pool (Weismann 2008: 29). Weismann's notion of a structurally impermeable 'barrier' between *germen* and *soma*, genetic memory and the memory of the individual nervous system, maps onto Jacob's theory of the inert genetic 'programme' and accounts, according to Stiegler, for 'the negentropic strength of the non-human life-form' (L, 67–8/63–4). Following Leroi-Gourhan, however, Stiegler also posits a third or 'tertiary' kind of memory, externalised in heritable technologies, which is characteristic of humans and 'can no longer be inscribed within the [adaptationist] framework of neo-Darwinism' (TT3, 206/301). In addition to heritable but not cumulable genetic memory and accumulable but not heritable somatic memory (which Stiegler also calls epigenetic memory),[2] there is, in the case of humans, 'epiphylogenetic' or material memory, sedimented in the technical artefacts of society. It is through the invention of epiphylogenetic memory that human evolution departs from the restrictiveness of adaptation by natural selection.

Stiegler follows but also sees himself as correcting Leroi-Gourhan on this point. The latter, he suggests, failed to grasp the full extent of the relation between the use of tools to create external memory and the emergence of 'reflexive' human intelligence (TT1, 178–9/186–7). For Stiegler, our interiority, which is to say the inner life of the mind, does not precede tool-use, but is rather constituted by it, performatively enacted though technics: 'man invents himself in technics by inventing the tool – by "externalising" himself techno-logically. [. . .] The interior is invented through this movement: it cannot precede it' (TT1, 141–2/152). Technology gives rise to a new relationship between the organism and its environment, according to which the 'who' ek-sists and is performed through the 'what' it employs to interact with its surroundings. Instead of responding to environmental conditions by just living or dying, depending on fitness, organisms evolve technically when they use tools to anticipate and ward off threats, by transforming their environment. Humans are able to draw on a body of past experience, *savoir vivre* and *savoir faire* externalised in the technologies of human society, and this means that acquired characteristics do not simply die alongside the organisms that acquired them. 'The most archaic technical evolution is already no longer "genetically programmed"' (TT1, 135/145), but rather marks a rupture with the purely biological understanding of natural selection, survival and adaptation.

> The phenomenon of life *qua* Dasein becomes singular in the history of the living being to the extent that, for Dasein, the epigenetic layer of life, far from being lost with the living being when it dies, conserves and sediments itself, passes itself down in the order of survival [*survivance*], and to posterity as a gift as well as a debt [. . .] not as a 'programme' in the *quasi*-biological determinist sense [. . .] This epigenetic sedimentation, a memorisation of what has happened, is what we call the past, what we shall name the *epiphylogenesis* of man, meaning the conservation, accumulation and sedimentation of successive epigeneses. Epiphylogenesis is a break with pure life, in that in the latter,

epigenesis is precisely what is not conserved ('the programme cannot receive lessons from experience'). (TT1, 140/151)

Beyond the '*quasi*-determinism' of Jacob's biological programme, in the passage from the genetic to the non-genetic, we see the advent of a new kind of *technical* programme. Genetic memory is immune to experience and somatic ('epigenetic') memory is perishable, but humans' technical memory is neither of these: 'the epigenetic experience of an animal is lost to the species when the animal dies, while in a life proceeding by means other than life, the being's experience, registered in the tool (in the object), becomes transmissible and cumulative: thus arises the possibility of *heritage*' (TT2, 4/12).

Adoption and adaptation

The temporality of the living is a negentropic temporality, which is to say a temporality that fights against disorder and which already involves a *différance* [. . .] a structure that differs from and attempts to defer [*différer*] entropy, that fights against entropy. This is life. (TDI, 155)

The difference between technical and biological evolution is between two regimes of differentiation, two ways of staving off entropy. Human life is differentiated not at the level of the gene, but through technics. 'The *who* is not differentiated like other living beings; it is [. . .] differentiated by the non-living [. . .] by organised but inorganic matter, the *what*' (TT1, 154–5/163–4). The life of the mind, or spirit, in other words, is inscribed, or externalised, in technical objects and constantly rewritten through the internalisation of these objects. Humans exist as *différés*, differed and deferred, in a state of suspension, constitutively open to being technologically rewritten in the future (TT1, 231/237). 'What we call "human", the technical, is but suspensions – deriving from [. . .] the default of origin', the absence of originary interiority (TT2, 160/185). We experience this suspension in technics as anticipation and ultimately desire. 'The being who defers [*diffère*] by putting off until later anticipates: to anticipate always means to differ and defer [*différer*]' (TT1, 231/237).

Non-human life evolves through adaptation by natural selection, but humans evolve *technically*, using tools to anticipate possibilities not contained within any biological programme (TT1, 151/160–1; TT3, 90/142). Rather than begin again from scratch with every generation, we are born into a technical symbolic order whose past we adopt as our own through participation in tradition. Stiegler returns to Leroi-Gourhan in *Technics and Time, 3*, borrowing from the latter a concept of 'adoption' to formulate the selection through tradition by which we inherit the technical *savoir vivre* accumulated by society. The extent of our adoption of technics, namely our access to the technologies through which society operates, is what determines our ability to participate in the construction of the institutions and values in which our artificial environment consists. Given the overwhelming inertness of our genetic inheritance, it is also through the adoption of different technics (clothes, books, and so on) that we differentiate (or 'transindividuate') ourselves within a group. By enabling the active construction of a future, an 'à-venir' that is not simply the passive acceptance of what one 'becomes' ('devient'), adoption is what makes technical evolution irreducible to Darwinian adaptation: '*adoption is not a simple adaptation to becoming* [devenir], but precisely its *projective transformation* into a possible future [*avenir*], as the implementation of a criterion that has been invented' (TT3, 175/260). The invention of new environments through 'non-living' technics marks the projection of life into the future, leading Stiegler to state that 'human time is negentropic in an extreme sense'. Created futures are nonetheless fragile, susceptible to regressing, or collapsing back, into the entropic becoming of the physical universe (L, 42–3/20; TDI, 155).

Stiegler is not always consistent in construing 'becoming' as negative, but it is worth pausing to note his choice of the term, here. The contemporary concept of becoming comes most notably from Deleuze, who positively valorises its role in a philosophy that rejects the reduction of life to biology. Deleuze follows Nietzsche in elaborating an 'ontology of selection' that critics including Stiegler's daughter have recognised as constituting a critique of passive adaptationism (Deleuze 2006: 62, 71–2/69, 81–2; Stiegler 2001: 96–107). Deleuzian concepts like 'becoming-animal' and 'becoming-inhuman' are central to this, expressing active creations of vitality (Deleuze and Guattari 2004b: 36–8/47–8). Stiegler recognises this aspect of Deleuze (EC, 81–92 [§§18–20]), but in his descriptions of becoming as entropic, there is perhaps also a subtle criticism of a conception of life that risks being read as closer to entropy, the breakdown of order, than to negentropy, the organisation of resistance to collapse (see, for example, DD3, 28). There is already a hint of this in Deleuze's affirmation of life as an impersonal but singular 'event', a 'body without organs' that is unencumbered by the (negentropic) 'organisation' of life into individuated identities; likewise in his apparent advocacy of the breakdown, or 'deterritorialisation', of structure. The problem, Stiegler suggests elsewhere, is that Deleuze's non-biological vitalism does not go far enough – does not go so far as to recognise the fragility of human life, constituted as it is through technics (MS, 104, 107/108, 110). Deleuze risks over-identifying the human with the anthropocentric, metaphysical assumption of biological privilege that he criticises as an 'imprisonment of life' (Deleuze 1999: 109/139). And by conflating humanity with a regime of repression, to be transgressed, he is blinded to the difference between the (negentropic) destruction of harmful forms of order and the (entropic) destruction of (negentropic) structures of society. By offering no more than notional 'resistance' to their collapse, Deleuze and his generation underestimate the importance of *constructing* the artificial environments in which we technically *ek-sist*. The philosophy of *übermenschlich* 'becoming-inhuman' is liable to give way to dehumanising bestialisation, lending itself to co-optation by a neo-Darwinist ideology of adaptation. Referring to Deleuze, Foucault, Derrida and Lyotard, Stiegler writes:

> The thought elaborated by these imposing personalities in France, between 1960 and 2000, has left their heirs unarmed, and a certain way of inheriting them has doubtless led to a genuine sterilisation of thought itself – often giving the impression of rationalising and legitimating the *renunciation of thinking an alternative*, for example, by positing that there is *no* alternative to the *state of affairs* that leads to universal unreason, other than 'resistance' to a kind of *inevitable fate of stupidity* and *performance*, imposed on us as a new regime in which knowledge has become an 'information commodity'. (EC, 133 [§30])

Read from a particular angle – inherited in a certain way – poststructuralism legitimates the impression of there being no alternative to an economic regime that promotes dehumanising values of performance and bestialising stupidity. Stiegler's use of the phrase 'no alternative' is already revealing, deliberately echoing a trope of the nineteenth-century evangelist of social Darwinism,

Herbert Spencer, who argued that 'there is no alternative' to an acceptance of Darwin, a recognition of the 'progressive, unbroken evolution' of 'Life under all its forms' (1855: 578 [§197]). More significant, for Stiegler, is the phrase's entry into common parlance in the 1980s, when it re-emerged as the acronym, TINA, to serve as a slogan of Reagan and Thatcher, who used it to insist that 'there is no alternative' to marketisation. On the contrary, Stiegler argues, humans' ability to invent themselves through technics means that 'there are lots of alternatives' (WML, 198fr [§68]).

Tristes entropiques: the performance of adaptation

What if we were already no longer humans? (TT1, 136/146)

According to Stiegler, adoption 'has *strictly nothing to do* with a process of *adaptation*. Adaptation is the factual state of animals and animalised human beings – which is to say, of slaves' (TT3, 168/250). Having previously affirmed that 'the tracing of any simple boundary between humanity and animality must be seriously called into question' (TT1, 151/161), he states elsewhere that animals, unlike humans, are incapable of epiphylogenesis, bound by genetics in a way that prevents them from creating a future beyond adaptation: 'An animal can be no more than unperturbably animal. [. . .] There where an animal is unperturbable, a man must always go beyond himself.' Animals cannot become human, but 'there is in man the constant possibility and temptation to regress to animality' (CE2, 58, 57 [§6]).

On first reading, the claim appears needlessly provocative, moreover undeconstructive, inviting the suspicion that Stiegler upholds an essentialist difference in kind between (technical) man and (biological) animal. But that would be to misread the descriptiveness of his definition of man and human, terms that he uses interchangeably. A being is human, for Stiegler, in so far as it evolves epiphylogenetically; in so far as it can create itself a future through the inheritance and adoption of tool use. Far from elevating *homo sapiens* above the rest of the animal kingdom, his point is that humanity has no essential basis in biology; it is a behaviour potentially performable by all kinds of life. 'The most terrifying thing would be for *The* Human to exist. It does not exist. [. . .] Humans exist' (TT2, 162/187), multiple degrees of humanity, incorporating not just us, but also the great apes thought to inherit cultures of tool use (see, for example, Luncz 2012: 4), and any other being exhibiting evidence of culture, the inheritance of acquired experience. The extent of one's humanity is determined by one's capacity for self-invention, which, in the case of *homo sapiens*, is effectively ever-present. However much one might be reduced to the status of 'animal' or 'inhuman', we are never less than 'non-inhuman beings' (PS, 183/326).

Rather than a question of animality, regression, in this respect, pertains to 'bestiality', which sets in when the desire and motivation to elevate (*s'élever*) ourselves above animality is short-circuited, resigning us to mere adaptation. 'Bestialisation is what causes entropy to set in, what destroys the negentropy that constitutes life and, in particular, human life as an affirmation of the singularity of each individual' (CE2, 58–9 [§6]). The entropy in question is not 'natural', or biological, but technical and moreover industrially produced (EHP, 101). Stiegler attributes it to a model of social and economic Darwinism that manufactures homogeneity, destroying human vitality by liquidating the epiphylogentic bases of traditional *savoir faire* – the technical equivalent of biodiversity – and suppressing the difference between technical and biological evolution (CE2, 64 [§8]). The marketing technologies of consumerism naturalise adaptation, short-circuiting adoption by promoting technologies that inculcate a passive, lobotomised mindset, and this disinclines us to do anything more than acquiesce in the demands of the market. Contemporary capitalism thus amounts to a 'règne de la bêtise', a seemingly inescapable reign of bestial stupidity (EC, 108–9 [§25]), which leaves us locked in an interminable present with neither past nor future, unable to envisage an alternative to what is, at best, a resigned and ironic resistance to adaptation. Without the traditions whose adoption served as the condition of the possibility of anticipation, our ability to invent the future has been undercut by 'a *systemic stupidity* [bêtise] *that structurally prevents the reconstitution of a long-term horizon*' (CPE, 5/12). Stiegler goes so far as to identify bestialisation with the Marxist concept of proletarianisation: 'to adapt is to proletarianise, which is to say, to deprive of knowledge those who must submit to that to which they adapt' (WML, 254fr [§84]).

In the technical-evolutionary context of *adoptive* selection, the concept of performance, or rather performativity, refers to the invention of artificial environments through technics. But this sense of performativity, Stiegler argues, has been occulted

and replaced by an *'adaptationist ideology of performance'*
(CE2, 60 [§7]), which 'aims at the imposition of adaptation in
order to short-circuit adoption' (WML, 230, 240fr [§§77, 80]).
A capitalist discourse of adaptationism has elevated a pseudo-
scientific, economic concept of performance to the status of a
new 'criterion of selection' through which to assess the 'fitness'
of members of society, bringing about a climate of survival of the
fittest in the epiphylogenetic sphere of 'la vie de l'esprit' (CE2,
31 [§1]). Instead of invention, performance has come to mean
the optimisation of efficiency, the maximisation of output relative to input. The term is prominently employed in Jean-François
Lyotard's *The Postmodern Condition* (1979), where it designates
the self-adjustments a system makes to combat entropy. Lyotard
describes performance as the only goal still left to be pursued in
an age where, profit aside, universal values no longer hold sway
(Lyotard 1984: 62–4/31–3). Stiegler, by contrast, emphasises its
relation to the market – 'the injunction to be performing, which
is to say, quantifiable and utilisable by industry' (CE2, 54 [§5]) –
and criticises the identification of performance with negentropy. In
Constituer l'Europe 1, he traces the term's contemporary usage to
nineteenth-century horse-racing, from where, following its application to cars and machinery, it passes into industry via Frederick
Winslow Taylor's theory of the 'scientific management' of labour.
The mechanisation of production enables human labour to be
measured against that of 'the machine-tool, to the performance
of which the proletarian must *adapt* their own' (CE2, 50–1 [§4]).
Since the twentieth century, such quantification has been performed not just on workers, but also and above all on consumers,
their attention relentlessly mined and proletarianised by marketing technologies whose function is to channel desire into 'new'
possibilities of expenditure (WML, 210fr [§71]). The result is
what Stiegler, following Paul Valéry, calls a catastrophic 'crise de
l'esprit', a systemic breakdown of the life that elevates humanity
over animality. The all-too tangible symptoms of this crisis range
from widespread demotivation, depression and even 'the loss of the
feeling of existing' (WML, 11–16fr [Introduction]). Exemplified in
the rise of the pharmaceutical industry, the adaptationist response
to these symptoms includes the prescription of drugs that enable
performance to be maintained irrespective of one's underlying
vitality. So-called attention-deficiency medications offer everyday
functionality in the school or workplace, albeit anaesthetised, and

at the cost of reducing the life of the psyche to a chemically regu-
lable, 'purely biological function' (PS, 101/184).

Stiegler's concern over socio-economic Darwinism finds support
in the work of the economic sociologists, Luc Boltanski and Eve
Chiapello, who have written on the spread from business to
society of an 'unduly mechanistic' mentality of ' "adapt" or die'.
Their *New Spirit of Capitalism* (1999) documents a tendency,
since the 1980s, to peg employment to 'aptitude', the French
term used to translate Darwin's 'fitness', with those deemed 'least
"adaptable" ' routinely excluded from an increasingly competitive
labour force, swept aside by a selection process deemed so trans-
parently 'natural' that no-one is any longer seen to be doing the
selecting (2005: 235/345). The rebranding of economics in the discourse of evolutionary biology has served to 'naturalise' dramatic
changes in the social fabric as seemingly unavoidable 'mutations'
to which society has no choice but to adapt (2005: 193–4/300–3;
xii, 'Preface to the English Edition').

Boltanski and Chiapello go further still, making explicit a link
that Stiegler advances only more tentatively, between post-1968
economic reform and the 'artist critique' of poststructuralism.
They argue that poststructuralist philosophy, with its affirma-
tions of desire, play and the overturning of purportedly repressive structures, 'has found itself recuperated and turned into
profit by capitalism', rendered not just critically impotent, but
moreover firmly complicit in the 'disorganisation of production'
and the 'deconstruction of the world of work' (2005: 173, 217,
323–4/268, 317, 460). Despite being amongst the first (with
Félix Guattari) to criticise the exploitations of consumerism,
Deleuze, in particular, is depicted, or reterritorialised, as a key
theorist of 'adaptability' in the labour market, his concept of
the 'nomad' providing the blueprint for the postmodern worker:
mobile and unencumbered by traditional social ties, '(a vocation,
a profession, marriage, etc.)', which is to say, flexible and fit for
working to short-term contracts. In a series of references to a
commodified and soundbitten Nietzsche, which do not survive
in the English translation, Boltanski and Chiapello characterise
the new economy in terms of 'hommes légers' and 'la légèreté des
êtres' ('lighthearted men' and the 'lightness of beings'), who, for
all their supposed gaiety, find the lightness of their being increasingly unbearable, on account of its precariousness (2005: 122–4,
224/198–201, 328–9). The language is that of the *Übermensch*,

or at least Zarathustra, the dancer over life whose laughter alone suffices to kill his 'arch-enemy, the spirit of gravity' (Nietzsche 1969: 215). The ideal worker, by implication, is one who performs his or her fitness by laughing off and even actively affirming the instability of performance-related project work. What we see here is not just the requirement to adapt, but moreover to enjoy one's adaptation, to embrace it as an opportunity for 'becoming', in the Deleuzian sense of the term. Withstanding the pressure of selection to thrive in our economic environment is henceforth not just a matter of 'adapt or die', but of *adapt and smile or die!*, to borrow from the title of a book, by Barbara Ehrenreich, on the pernicious culture of the positive mental attitude.[3] The ideological imperative to enjoy one's adaptation conveniently fits consumer capitalism's channelling of redemption into the purchase and consumption of short-term pleasure.

What Boltanski and Chiapello describe as the resulting 'crisis', 'the end of critique', recalls Stiegler's claim, cited above, that a certain inheritance of poststructuralism has rendered it complicit in the market, resulting in 'a genuine sterilisation of thought' (EC, 133 [§29]). And yet, in the third volume of *Disbelief and Discredit*, Stiegler is highly critical of Boltanski and Chiapello, taking them to task for the superficiality of their readings of *la pensée '68*. There is nonetheless something ambivalent in his criticism, which consists in contesting their positive assessment of May '68 as a 'renewal', albeit fleeting, of the spirit of capitalism, as opposed to an instance of 'bêtise' and 'the loss of mind, or spirit' (DD3, 52). Far from a revolutionary liberation of repressed desire, he argues that 'the upheavals of 1968 are the first political, economic and social symptoms of a loss of spirit', early evidence of the destructive, immiserating mindlessness of hyperindustrial (consumer) capitalism (DD3, 11–12). The same symptoms manifest in the thought of that era. Deleuze and Guattari's joyous affirmation of life in terms of 'desiring-machines', he suggests, are a product of the capitalism they set out to criticise. Beneath the veneer of emancipation, what plays out, in their attack on the traditional institutions of authority, is a liquidation of the adoption processes that invent us as humans, by sublimating libido into anticipatory desire (DD3, 28). Stiegler appears torn between defending the poststructuralist tradition he has adopted as his own and putting forward a variation of the very criticism he is looking to resist. In an age of catastrophic social entropy, the ethos of poststructuralism becomes the light-headed, positive spin on a resurgent economic Darwinism.

Conclusion: the promise of the human

> The *Übermensch* is [. . .] the man fired by [*chargé de*] animals [. . .] It is the man fired by rocks, too (there where silicone reigns). It is the advent of a new form, neither God nor man, and which one can only hope will be no worse than the two that came before it. (Deleuze 1999: 109–10/140–1)

In the opening pages of *The Order of Things*, Michel Foucault famously announces 'that man is only a recent invention, a figure not yet two centuries old, a new fold in our knowledge, and that he will disappear again as soon as that knowledge has discovered a new form' (Foucault 2001a: xxv/15). More than two million years separate Foucault's version of the invention – that of man as both category and ground of modern epistemology – from what Stiegler identifies as the epiphylogenesis, or technical evolution, of the first humans. The difference is much smaller when it comes to their diagnoses of man's passing, though they diverge in their respective attitudes toward it. *The Order of Things* concludes on the affirmation of a disappearance that is already under way. As 'anthropology' dissolves into new 'folds' of knowledge, Foucault heralds 'the explosion of man's face' from an *übermenschlich* laughter that fills the space once occupied by a now dead god (2001a: 420/395–6). For Stiegler, by contrast, there is something *bête* in this laughter.

   The biological organism simply lives or dies depending on adaptive fit, but humans transcend adaptation. What makes life human is its constitutive openness to being rewritten by technics, and it is by adopting technics that we anticipate change, stave off the need for survival of the fittest. The human, in this respect, is not a dubiously privileged, essentialised biological entity, to be passed over on the way to something 'better'. It is the performative enactment of what Jacques Derrida would call an (uninstantiable) promise, projected into the future and irreducible to the present (Derrida 1995a: 15, 25). The problem with promises is that they can easily be broken. They are also liable to fade from memory, become meaningless, if they are not repeatedly reiterated, or reinscribed in experience; if the acquired experience that serves as the condition of their possibility is liquidated (Moore 2012: 277–8). The loss

of the ability to construct the promise of humanity is what proletarianisation, the regression to bestiality, consists in. And Stiegler's
concern is that his philosophical precursors lend themselves to this
process, by glorifying the collapse, rather than the construction, of
the human.

Capitalism short-circuits the invention of the human, leaving
us with no alternative but to adapt to the demands of the market.
Following Stiegler, Boltanski and Chiapello, one might wonder
whether, in the desire to become *über*-human, and in thrall to the
fantasy of our *Übermenschlichkeit*, we haven't helped to perform,
or invent, a world to which humans are increasingly maladapted,
and in which only the *Übermensch* would pass as 'fit'. Faced with
the threat of being eclipsed by science, can philosophy do more
than resist, and laugh off, its own entropy? We began with the
question of the 'ridicule' of French philosophy, posed by Michel
Houellebecq, whose novels, incidentally, are all about the fate
of worn out and broken, would-be Zarathustras (Moore 2011).
Stiegler's solution to this ridicule is to shift philosophy away from
the celebration of the end of the human, and onto the earnest construction of a future for humanity.

Notes

1. Houellebecq 2001: 376/314
2. Stiegler's use of the term 'epigenetics' to mean a mechanism that
   functions 'on top of', or in addition to, genetics, is inherited from
   Leroi-Gourhan and should not be confused with the term's use in
   contemporary evolutionary biology. Research in the nascent field
   of epigenetics has shown that some genes switch themselves on and
   off in response to environmental changes, and transmit 'heritable
   epimutations' to subsequent generations (Haig 2007: 421). Debate
   is split over whether these intergenerational switches are a matter of
   undirected, random adaptations that just happen to fit the new environment, or whether they amount to instances of anticipation and
   learning from experience (2007: 424–6). Either way, epigenetic evolution would be no more than a weaker version of the environmental
   anticipation that Stiegler attributes to technical evolution.
3. *Smile or Die: How Positive Thinking Fooled America and the World*
   (London: Granta, 2009). The original American title is *Bright-Sided:
   How the Relentless Promotion of Positive Thinking Has Undermined
   America*.

2

The Prehistory of Technology: On the Contribution of Leroi-Gourhan

Christopher Johnson

Bernard Stiegler's *Technics and Time* can in many respects be seen as a thematic continuation of the philosophical programme initiated some three decades earlier in Derrida's *Of Grammatology* (1967).[1] Put simply, where the governing theme of *Grammatology* was the question of the 'repression' of writing in Western metaphysics, Stiegler's text reformulates this question in terms of the repression of technology, undertaking a series of symptomatic readings of the history of that repression. A second, and important, point of intellectual continuity between the two texts is their reference to the work of the prehistorian and anthropologist André Leroi-Gourhan (1911–86). Like Derrida, Stiegler sees Leroi-Gourhan's palaeoanthropology as providing an essential starting point for a non-metaphysical reflection on the nature of the human. Unlike Derrida, whose treatment of Leroi-Gourhan in *Of Grammatology* is relatively brief and selective, Stiegler devotes substantial passages of the first and second volumes of *Technics and Time* to commentary of Leroi-Gourhan's texts. It is through this extended dialogue between philosophy and anthropology in *Technics and Time* that a number of key concepts of Stiegler's early philosophy of technology emerge and are developed: the idea of a 'maieutic' relationship between the human and the technological, the *who* and the *what*; the concept of 'tertiary memory'; the notions of advance and delay. At the same time, the convergent dialogue between Stiegler and Leroi-Gourhan reaches a critical limit in the first volume of *Technics and Time* as Stiegler begins to question what might be termed Leroi-Gourhan's delimitations of the pre- or proto-human. The intention of the following analysis is to examine these points of intellectual convergence and divergence, asking to what extent Stiegler's philosophical critique of Leroi-Gourhan may be missing its mark, and asking whether certain points of convergence between the two thinkers may not in their turn be questioned and subjected to another form of critique.

The contribution of Leroi-Gourhan's anthropology to Stiegler's philosophy of technology lies first and foremost in its systematically materialist approach to the definition of the human. From an early stage in his career, Leroi-Gourhan was questioning what he considered to be the intellectualist bias of the anthropology deriving from the sociological tradition of Durkheim and Mauss, in which the role attributed to technology in the study of human society was normally a subordinate one (Leroi-Gourhan 1993: 148/210–11, vol. 1). Stiegler's initial engagement with Leroi-Gourhan in the first chapter of *Technics and Time, 1* is a reading of his pre-war anthropology of technology as set out in the two volumes of *Évolution et techniques*, published in 1943 and 1945. Much of *Évolution et techniques* consists of extensive taxonomic descriptions of the elementary transactions of humans with different categories of matter, but there is also a self-reflexive methodological dimension to Leroi-Gourhan's investigation which brings it squarely into the domain of philosophy. The key concept here is that of the technical tendency (*tendance technique*), a concept which will be central to Stiegler's subsequent analysis of the dynamics of technological development in *Technics and Time*. For Leroi-Gourhan, the technical tendency refers to the technological determinism which dictates that humans will engage with their external environment (*milieu*) in predictable and convergent ways, and that the aggregate tendency of technological evolution will be towards an increasingly effective engagement with that environment.[2]

For Stiegler, the interest of Leroi-Gourhan's concept of technical tendency lies in its non-anthropological perspective on the evolution and history of technology. In Leroi-Gourhan's investigations of what Stiegler terms the 'coupling' of the human and material, traditional diffusionist explanations of technical development are consistently bracketed out in favour of the idea of technological convergence – it is often impossible to reconstruct the origin and the different routes of transmission of technical complexes. This does not negate the historical fact of technical development through the communication and exchange of technologies between human groups, but rather situates the dynamic for technical development at a higher level of determination. Much like the phenomenon of convergence in biological evolution, where genetically distinct species may under similar environmental conditions assume near-identical forms, the range of functional transactions between the human and the material is finite, so that similar technical complexes may emerge independently in geographically distinct human groups. As Stiegler points out, Leroi-Gourhan's reference to the biological in *Évolution et techniques* is both metaphorical and more than metaphorical: technical evolution is both *analogous* to biological evolution and, in its coupling of the human and the material, a *continuation* of biological evolution (TT1, 48–9/63). This equation between the biological and the technical, what Stiegler terms the 'zootechnological relation of the human to matter' (TT1, 49/63), will be an important component of his analysis of Leroi-Gourhan in the third chapter of *Technics and Time, 1*. More generally, Leroi-Gourhan's definition of the technical tendency helps Stiegler to conceptualise the *autonomy* of technical development in relation to the human. In contrast with the anthropocentric perspective of a human invention of technology, locating the 'origin' of this or that technical complex in this or that geographical group, Leroi-Gourhan argues for the differential 'materialisation' (Stiegler uses the Simondian term 'concretisation') of the technical tendency in ethnically distinct human groups. For Stiegler, the interest of this determinism of technological form and function is its ultimate indifference to the ethnic: it is both constitutive of the difference between cultures and ultimately the vector of the dissolution of cultural difference in the modern world (TT1, 64–5/79; see also TT2, 76–7/94–5).

Combined with Stiegler's readings of other theories of technology (Gille, Simondon, Heidegger) in the first chapter of *Technics and Time, 1*, Leroi-Gourhan's pre-war anthropology of technology therefore provides Stiegler with a model of technical evolution that will inform his more general reflections on the state of technology in the contemporary world. Equally importantly, Leroi-Gourhan's post-war work on the prehistory of technology can be said to provide a natural-historical grounding for and validation of Stiegler's thesis concerning the co-determination of the human and the technical. In what is structurally the central chapter of *Technics and Time, 1*, Stiegler undertakes a detailed reading of Leroi-Gourhan's *Gesture and Speech* (1964–5), an exemplary text whose intellectual legacy, he argues, has still not been properly assumed in either palaeoanthropology or philosophy (TT1, 84/97).

As Stiegler indicates, *Gesture and Speech* opens with a critique of traditional metaphysical definitions of the human, such as that of Rousseau's *Second Discourse*, which present an original humanity fully formed in body and mind but lacking both the 'arts' of culture and the structures of society. Rousseau's speculative reconstruction of the origins and development of humanity is a 'transcendental' anthropology, which has to begin by setting aside the 'facts' in order to explain the passage from the state of nature to the state of culture (TT1, 84/97). The palaeoanthropology of *Gesture and Speech*, by contrast, starts from the facts of evolutionary sequences. Importantly for Stiegler, Leroi-Gourhan begins his narrative of human evolution with the pre-human, in the extended series of vertebrate forms, some of which will eventually converge on the human. This places the animal before the human, but it also places the anatomical before the cognitive. Leroi-Gourhan treats the functional anatomies of animal forms as different *engineering* solutions to the vital requirements of mobility and prehension in different environmental conditions. At each stage of a given evolutionary sequence, a 'balance' or equilibrium is achieved between organs dedicated to locomotion and the forward-facing organs dedicated to orientation and prehension – what Leroi-Gourhan terms the 'anterior field' (*champ antérieur*). Leroi-Gourhan emphasises that the development of nervous systems to 'control' the operations of the anterior field is secondary to the development of the skeleton, the mechanical infrastructure which articulates movement. As Stiegler comments, quoting Leroi-Gourhan, 'mobility, rather than intelligence, is the "significant feature"' (TT1, 146/156). According to this interpretation, the evolutionary singularity which will distinguish the human from proximate animal forms such as the primates is not the brain but the feet: the emergence of full bipedalism permits a further 'liberation' of the anterior field, freeing the hands for more complex and mediated interaction with the material world. Thus, *Zinjanthropus boisei*, the earliest hominid form known at the time of *Gesture and Speech*, presents a braincase markedly inferior in volume to that of anatomically modern humans, but is fully bipedal and already in possession of rudimentary stone tools.[3] In the first volume of *Gesture and Speech*, Leroi-Gourhan will describe the successive stages leading from *Zinjanthropus* to Neanderthal, the most advanced hominid form before *Homo sapiens*.[4] What the palaeontological and archaeological records demonstrate is a general synchrony between technical and cognitive evolution, a complexification in tool stereotypes accompanied by a growth in brain size and in particular expansion of the cortex. For Stiegler, the significant feature of this evolution is that it is not the product of a fully-formed human intelligence: it is not the human mind that invents technology, rather the human mind is invented with and through technology in a two-way process, a 'structural coupling' between the human and material which he terms an 'instrumental maieutics' (TT1, 158/167). Temporally, Stiegler formulates this circular relationship in terms of 'advance' and 'delay': there is an advance of the material (the anatomical and the mechanical) over the mental (nervous system, brain, cortex), and a delay, a retardation of the human (mind and body) in relation to its 'externalisations' in the instrumental world (TT1, 145/155).

Stiegler describes the evolutionary emergence of human cognition as a succession of 'mirror stages' in which the human achieves self-reflexive consciousness through its manual engagement with the material world: the cortex is (metaphorically) reflected in the piece of flint. But the mirror of technology is also a memory. The artefact endures as a trace, a record of a process of manufacture external to the human agent: 'Flint is the first reflective memory, the first mirror' (TT1, 142/152).[5] Leroi-Gourhan refers to the process of manufacture itself as the operational sequence (*chaîne opératoire*), the structured chain or sequence of actions necessary to extract, for example, a specific instrument (tool or weapon) from the raw material of the block of flint. The operational sequence therefore presupposes a certain intentionality, a capacity for anticipation in the agent of technology, from the earliest forms of human intelligence onwards.[6] As Stiegler will later comment, the concept of the operational sequence allows Leroi-Gourhan to think of language and technics as co-emergent features of human cognition, both dependent on a process of abstraction and a 'syntax' of operations (TT1, 167/176). This concept also contributes to Stiegler's formulation of what will be a central component of his philosophy of technology in *Technics and Time*, the idea of 'tertiary memory'. Leroi-Gourhan describes the specificity of the human as residing in three categories or levels of memory. The first two levels, genetic and epigenetic, are the categories of memory shared with other biological species, the hereditary memory of genetic reproduction and, increasingly in mammals, the neurological memory of individual experience. The singularity of the human, however, resides in the type of extra-individual (shared, social) memory deposited in the operational sequences of language and technology. Stiegler refers to the operation of this third level of memory as *epiphylogenesis*, an essentially cumulative process which in its turn exercises a powerful selective pressure on the biological human: 'It is in this sense that the *what* [the techno-logical] invents the *who* [the human] just as much as it is invented by it' (TT1, 177/185).

Stiegler's reading of *Gesture and Speech* in *Technics and Time, 1* is a generally convergent reading, to the extent that the dialogue it sets up between philosophy and anthropology is a mutually confirming one. From what is probably the most philosophical of Leroi-Gourhan's texts, Stiegler extracts a series of concepts (liberation, externalisation, operational sequence, etc.) and translates these into the terms of his more general philosophy of technology: instrumental maieutics, advance/delay, epiphylogenesis, the *who/what*, etc. The strength of this reading is that it confirms and continues what is the fundamental materialism of Leroi-Gourhan's account of human origins, presenting human evolution as the synthetic co-emergence and co-determination of mind, technics, language and memory. However, what I find most interesting in this reading is the point at which it diverges from its reference text, where Stiegler begins a critique of what he considers to be the residual essentialism of Leroi-Gourhan's thought. This has to do with the different *stages* of the human described by Leroi-Gourhan in *Gesture and Speech*. Whereas Leroi-Gourhan characterises the evolutionary sequence from *Zinjanthropus* to Neanderthal as a progressive 'humanisation' of the human (Stiegler uses the more conventional term 'hominisation'), he sees the appearance of *Homo sapiens* as marking a qualitative shift in this evolution, in which a species whose intelligence has been geared primarily to the material requirements of subsistence and survival begins to show signs of another type of intelligence, a spiritual and creative intelligence not immediately dependent on the technical intelligence of its origins. For Stiegler, this introduction of a 'second origin' of the human is logically in contradiction with the remainder of Leroi-Gourhan's demonstration, and is characteristic of the kind of metaphysical humanism Leroi-Gourhan had himself critiqued in the introduction of *Gesture and Speech*:

> The critique of Rousseau consisted in saying that the human is not a spiritual miracle that would be 'added' to a previously given body of

the primate. Now, with the second origin, something is 'added' to the technological: the *symbolic* or the *faculty of symbolisation*, without an understanding of its provenance. (TT1, 163/170)

Stiegler's critique of Leroi-Gourhan could be described as classically deconstructive, in many respects reminiscent of Derrida's treatment of Lévi-Strauss or Rousseau in *Of Grammatology*. He warns us that *'the greatest vigilance with respect to oppositions is called for – even if – and nothing is more difficult – the contestation of oppositions must not eliminate the genetics of differences'* (TT1, 163/172). The 'opposition' in question is Leroi-Gourhan's distinction between *Homo faber* and *Homo sapiens*, between technical intelligence and symbolic intelligence, and Stiegler is certainly right to alert us to the dangers of such essentialising oppositions. However, first it is not entirely clear that Leroi-Gourhan actually *opposes* these two types of intelligence. Second, as Stiegler himself seems to concede, the contesting of such oppositions may itself prevent us from being able to think the genetic, i.e., how one form of humanity may be followed or superseded by a different form of humanity. This qualification, I think, lies at the centre of the divergence between philosophy and anthropology – Stiegler and Leroi-Gourhan – as it develops in *Technics and Time, 1*, a divergence which could be said to derive from their differing perspectives on the question of time and on the question of 'symbolic' thinking.

As its title indicates, Stiegler's book is an investigation of the relationship between human technics and human time. This investigation is explicitly presented as a variation on 'the Heideggerian problematic of time' (TT1, 179/187), in which for Stiegler 'our most profound question is the technological rooting of all relation to time' (TT1, 135/146). The phenomenological-existential (lived) experience of past-present-future, the time of memory, consciousness and anticipation, is grounded in technics. It is technics which gives time, which is the origin and possibility of human time. As we have seen, in the convergent phase of Stiegler's reading of Leroi-Gourhan, palaeoanthropology provides the positive 'facts' and a conceptual framework for thinking such an origination. The concept of the operational sequence, based on the technical artefacts of the archaeological record, infers a capacity for conscious anticipation from the earliest of human forms. At the same time, these artefacts endure as an externalised residue of human intention and human action, and therefore as a replicable 'hard' memory, a 'mirror' of consciousness, as Stiegler describes it. The repetition of technological stereotypes from generation to generation creates an extra-individual, collective memory over and above that of the biological species.

Despite its analytical force, it could be argued that Stiegler's quasi-phenomenological framing of the question of technics and time generates a rather partial and somewhat reductive reading of prehistory, a reading which does not properly give the measure of evolutionary time. A symptom of this bias is Stiegler's relative indifference to Leroi-Gourhan's attempts to delineate different stages in the evolution of archaic humanity, and to explain the transitions between them. It is as if for Stiegler the philosopher, the principle of the technical constitution of intelligence is sufficient, and it is not necessary to reconstruct the stages through which different human forms may have developed in their convergence on anatomically modern humans. Here, I will take a sample of two short passages as examples of this attitude, the first referring to the beginnings of the human in *Zinjanthropus* and the second to its termination in *Homo sapiens*:

Either the human is human from the Zinjanthropian onward, in which case there is technico-intellectual intelligence as such in a single stroke, [which] means that there is anticipation in the full sense of the term [. . .] Or the Zinjanthropian is nothing but a prehominid who cannot anticipate, that is, who is not in time and who in no case accomplishes its future since it has none, no more than does the 'man of pure nature'. (TT1, 160/169)

On the basis of these specifications, which blur a too distinct boundary [*frontière*] between the different stages of the archaic human, Leroi-Gourhan introduces his major thesis on the last stage – the preponderant role played by society. [. . .] Must this mean that society was not there before? Certainly not. That there is a dynamic in which preponderances shift is obvious. But that boundaries [*frontières*] should be marked off in this dynamic is less satisfactory. Everything is there in a single stroke. Everything is differentiated in one coup, together. It is the *inorganic organisation of memory* that constitutes the essential element, the first coup engendering all the others and being transformed in transforming all the others in its wake. In this complex, the brain has in fact only a secondary role, in no case a preponderant one. (TT1, 174/182)

In the first passage Stiegler is responding to Leroi-Gourhan's attempt, in *Gesture and Speech*, to establish what might have been the cognitive differences between the earliest and most recent of human forms. In line with the synchronism of Leroi-Gourhan's analysis of the archaeological evidence – his correlation of successive anatomical forms with the complexification of tool stereotypes – he speculates that later human forms may be associated with a higher degree of reflexive intelligence and vice versa. On the one hand, Stiegler's response to Leroi-Gourhan is a logical one: it is difficult, perhaps impossible, to attribute levels of anticipation, to quantify or qualify degrees of intelligence and reflexivity in relation to different stages of hominid development. If one is searching for 'the essential element', as he puts it in the second passage above, then it is present from the beginning, in 'a single stroke', in the technical intelligence of *Zinjanthropus*. On the other hand, Stiegler's response is perhaps too starkly formulated in terms of the 'either-or', reducing Leroi-Gourhan's anthropology to the binary alternative of a *Zinjanthropus* who is either human or pre-human. This response is also limited, it could be argued, by its dependence on a certain type of philosophical discourse, a discourse which has a tendency – when confronted with the genetic, the historical and the developmental – to default to the explanatory mode of the 'always already'. This is quite foreign to Leroi-Gourhan's mode of thinking with respect to the process of evolution, a mode of thinking predicated on the 'both-and' rather than the 'either-or' and the 'not-quite-yet' rather than the 'always already': *Zinjanthropus* is both already essentially human (human = technics = language) and developmentally (anatomically, neurologically) not quite yet the same species of human as later hominid forms. As has been noted, Stiegler seems to be indifferent to, even impatient with – as is clear in the second passage quoted above – Leroi-Gourhan's attempts to reconstruct successive stages of the human, to establish boundaries (*frontières*) between them. As a result of this, his reading of Leroi-Gourhan seems temporally flat, characterised rhetorically by its repeated references to *Zinjanthropus*–Neanderthal as a single evolutionary sequence. This reading does not sufficiently take account of what the prehistorian or palaeontologist must take account of: the existence of different anatomical forms associated with different industries and the problem of explaining the transitions between them. In fact, despite the density of Stiegler's reading of *Gesture and Speech*, it is for the most part brief and elliptical on the detail of its science. While recognising, for example, that the evolution of the human brain configuration 'takes time' (TT1, 163/172), it is barely mentioned that this time is of the order of some two million years, nor that the transition to *Homo sapiens* takes place over a period that is relatively a fraction of this time. Similarly, Stiegler is not specific on the quantitative difference between the braincases of *Zinjanthropus* and Neanderthal, a difference of the order of approximately 1:3, increasing from 500 cm^3 in *Zinjanthropus* to more than 1500 cm^3 in Neanderthal. It is on the basis of these differences of evolutionary time and anatomical dimension that the palaeoanthropologist makes the evaluation that the philosopher appears reticent to make: that *Zinjanthropus* would have been cognitively less advanced than the hominid forms which followed it. By contrast, the most that Stiegler is able to say regarding such cognitive differences is that the 'technical and cerebral conditions' of *Zinjanthropus* are 'profoundly alien to us' (TT1, 173/181).

Equally important, at the qualitative level of expression, is what could be described as Stiegler's philosophical processing of Leroi-Gourhan's text, that is, the translation of his account of human cognitive evolution into the terms of a philosophy of *difference*. In both of the preceding passages, it is argued that the essential features of the human are present from the origin, in a 'single stroke'. In the second passage, it is qualified that 'Everything is differentiated in one coup, together'. And a few pages later, it is proposed that 'From the Zinjanthropian to the Neanderthal, cortex and tool are differentiated together, in one and the same movement' (TT1, 176/184). The deployment of this vocabulary of difference and differentiation will doubtless draw a certain category of reader into philosophically familiar terrain. Conceptually, it is consistent with Stiegler's attempts to think the temporality of technology, his formulation of a dynamics of 'advance' and 'delay' via the Derridean concept of *différance*. However, it can be asked to what extent this terminology does justice to the process of evolution 'From the Zinjanthropian to the Neanderthal', as it is described in Leroi-Gourhan's text. The concept of differentiation, it could be argued, provides only an approximate representation of the evolutionary process, which in Leroi-Gourhan's account takes place through a series of *additions* or *accretions*. For example, once the basic anatomical (skeletal) infrastructure of the human is in place (bipedalism, liberation of the hand), the growth in cranial capacity observed from *Zinjanthropus* onwards is not so much a process of differentiation as the progressive layering of brain functions, culminating in the full development of the pre-frontal cortex in *Homo sapiens*:

> Proceeding from the very general biological phenomenon of evolution employing earlier stages to serve as the active substratum for new, innovative ones, we have considered the evolution of the nervous system in terms of the addition of new cortical areas that led to the simultaneous emergence of technical motor function and language, and, later, to technicity controlled by mental processes [*une technicité hautement réfléchie*] and to figurative thought. (Leroi-Gourhan 1993: 251/56, vol. 2)

What is fascinating here is that the model informing Leroi-Gourhan's description of cognitive evolution as a process of accretion and stratification, a combination of the new and the residual, is a *technological* model. The section immediately preceding this passage evokes the electronic and cybernetic technology of the 'last twenty years', which has already achieved a comparatively high level of imitation of biological systems (*le vivant*), representing 'a synthesis of all previous stages'. These new automatic technologies force the biologist to view the living and the technical-artificial worlds as two parallel manifestations of the same process (1993: 250–1/55–6, vol. 2). If one were searching for an appropriate metadiscourse for the description of the processes of bio-neurological and technological evolution, it seems that the technical metaphor of *bricolage* would in fact provide a more effective means of conceptualising these processes than Stiegler's more abstract notion of differentiation. As the molecular biologist François Jacob puts it, evolution as *bricolage* is the 'constant re-use of the old in order to make the new' (Lévi-Strauss 2009: 50).

Stiegler's reservations concerning Leroi-Gourhan's delineations of different stages of human development are therefore symptomatic of a more general divergence between philosophy and anthropology in their modes of thinking about human and evolutionary time. Put simply, Stiegler's reading appears to be more concerned with the essential and the originary than with the developmental and the emergent. This in turn affects how he approaches the question of symbolic intelligence. Stiegler is particularly critical of Leroi-Gourhan's attribution to *Homo sapiens* of the capacity for a higher level of symbolic abstraction, and his

contrasting of the latter with the more 'concrete' forms of symbolisation that would have been available to preceding human types.
He argues that the idea of a 'concrete' symbol is a contradictory
concept, that – as Leroi-Gourhan himself admits elsewhere –
language is in essence and from the beginning based on a process
of abstraction (TT1, 168–9/176–7). While Stiegler's point is a
cogent one – Leroi-Gourhan's characterisation of different stages
of language evolution might indeed have been more carefully
formulated – nevertheless the question remains of the possibility and probability of a gradated evolution of language.[7] Again,
Stiegler's tendency is to argue that 'Everything is there in a single
stroke' (TT1, 174/182), a position which brings him close to Lévi-
Strauss's formulation in the *Introduction to Mauss* of a singular
and integral origin of language.[8] In line with his framing of the
question of technics and time within a 'Heideggerian problematic',
this originary symbolic intelligence is linked with the possibility of
an already properly human, *existential* relation to death:

> There is no [second origin] because technological differentiation pre-
> supposes full-fledged anticipation, at once operative and dynamic,
> from the Australanthropian onward, and such anticipation can only
> be a relation to death, which means that symbolic intellectuality must
> equally be already there. (TT1, 163/171–2)

The inference from what remains of *Zinjanthropus* (fragments of
bone and structured stone) of this package of human attributes
(anticipation, symbolic intelligence, consciousness of death) is
logically coherent to the extent that it is an extrapolation of what
we know of the human from our perspective as *Homo sapiens*.
However, it is not verifiable, and does nothing to explain the evolutionary distance between *Zinjanthropus* and Neanderthal. For
the prehistorian and anthropologist, of course, it is not a question
of the individual and existential, but of the collective and ritual
'relation to death', and there is no hard (durable) evidence of this
kind of behaviour until very late in the archaeological record, with
Neanderthal. Asserting that 'to clarify the meaning of "symbolic"
is to introduce the question of mortality', Stiegler goes on to quote
Leroi-Gourhan directly:

> Archaeological evidence of such activity – which goes beyond tech-
> nical motor function – [is] the earliest of an aesthetic or religious

character [*de caractère esthético-religieux*], [and] can be classified in two groups as reactions to death and reactions to shapes of an unusual or unexpected kind [*l'insolite de la forme*]. (TT1, 164/173)

Quite symptomatically, Stiegler's reading of Leroi-Gourhan following this quotation remains fixated on the question of death, and is silent on the second category of mental activity cited by Leroi-Gourhan, an omission which is repeated a few pages later in a further quotation from *Gesture and Speech* (TT1, 167–8/176). What Leroi-Gourhan is referring to here is the presence, relatively late in the archaeological record, of natural objects with no clear utilitarian function which were collected by our hominid predecessors, including Neanderthal. He takes this non-utilitarian behaviour as evidence of an emergent aesthetic sensibility, an advanced form of pattern recognition anticipating the later explosion of artistic activity observed in the Upper Palaeolithic period (Leroi-Gourhan 1993: 367–9/212–14, vol. 2). While Stiegler's critique of Leroi-Gourhan is restricted to the technical-linguistic dimension of symbolic intelligence – what Leroi-Gourhan would describe as a 'linear' mode of cognition based on the operational sequence – it is strangely elliptical on the wider, 'multidimensional' forms of symbolic activity evident in the archaeological record from Neanderthal onward.[9] This is curious, because for Leroi-Gourhan it is precisely the emergence of this wider capacity for symbolisation which explains the exponential growth of technological culture associated with *Homo sapiens*, what was referred to above as a 'highly reflexive level of technicity' (*une technicité hautement réfléchie*). Again in a passage quoted by Stiegler, Leroi-Gourhan argues that the new cognitive capacities evident in Palaeoanthropians (Neanderthal) act as both a 'counterbalance' and a 'stimulant' to technical intelligence (1993: 162/171). Whereas the evolution of the brain up to this point follows the process of externalisation and is determined by it, from this point onwards it is the brain which becomes the driving force.

To claim, as Stiegler does, that Leroi-Gourhan *opposes* technical intelligence and 'spiritual' or 'creative' intelligence, that he posits a unexplained 'leap' from one state or stage of the human to another, therefore does only partial justice to the narrative of human evolution as it is presented in *Gesture and Speech* – Stiegler's attribution of a 'second origin' of the human too readily conflates Leroi-Gourhan with Rousseau in this respect. Leroi-

Gourhan himself would doubtless agree with Stiegler that there is no second origin, but at the same time would argue that there are emergent properties at the stage of Neanderthal and *Homo sapiens* which set these species apart from previous human forms. 'Emergent' in this context does not mean *ex nihilo* – it is less the case of an evolutionary jump than a continuation of the process of stratification described above – evolution as the constant building upon the old in order to make the new – reaching definitive critical mass with *Homo sapiens*.[10] Nor does this emergence mean, again as argued by Stiegler, that previous human forms are presented by Leroi-Gourhan as being 'almost human' or less-than-human: the humanism of *Gesture and Speech* is much more inclusive than this simple opposition suggests.

If Stiegler's critique of Leroi-Gourhan in *Technics and Time, 1* is not a conclusive critique, there are also other aspects of his reading of *Gesture and Speech* which appear to suffer from a relative deficit of critique. This has to do with the wider moral-humanistic dimension of Leroi-Gourhan's exposition and its relationship with Stiegler's own particular philosophical programme, as he pursues it in the first two volumes of *Technics and Time*. As Stiegler notes, 'the end of the human cannot be investigated without investigating its origin' (TT1, 135/146). In fact, while Stiegler diverges from Leroi-Gourhan on the question of the 'second origin', he remains convergent with his evaluations concerning the evolution of technology following the bio-neurological stabilisation of the species with *Homo sapiens*. The dynamics of this evolution are played out between the three points of the individual, the technical and the social. At each stage of the evolutionary sequence leading from vertebrate to primate to hominid to *Homo sapiens*, Leroi-Gourhan describes what is essentially a process of accommodation or compromise between disparate forces: accommodation between different aspects of anatomical form and function in the skeleton; accommodation between the nervous system and the bodily articulations governing locomotion and prehension; accommodation between the neuro-motor areas of the brain and the fine articulations of the hand turned to the manipulation of matter. The operative term in all of these descriptions is 'balance' or 'equilibrium' (French: *équilibre*): the tendency of evolution as adaptation is towards the dynamic distribution of forces, a balancing out of form and function. The problem, and question, as the hominid sequence develops in the direction of an increasing externalisation of functions, is whether such an equilibrium can be maintained. Leroi-Gourhan's diagnosis, as his narrative moves into proto-historical and historical time, is that there is an increasing *disequilibrium*, a mismatch between the bio-neurological substrate of the human and the social and technological systems which have served as multipliers of its material transactions with the world. Translated into the terms of Stiegler's philosophy of technology, there is an 'advance' of the social and technological, and a chronic 'delay' of the human that is never properly compensated. Just as the individual human is progressively absorbed into the social organism, the latter is itself overtaken and dissolved by a technical tendency which is purely autonomous in its operation.

This is the tenor and orientation of Stiegler's reading of Leroi-Gourhan in the second volume of *Technics and Time*, a reading which seems in a number of ways to be critically less sharp than that of the first volume. Much of this reading consists of an extended commentary and quotation of *Gesture and Speech* which simply confirms Stiegler's own diagnosis of the 'disorientation' of contemporary existence. Paradoxically, the critical force which the philosopher had previously applied to the question of the 'second origin' is not applied to what is arguably a more problematic aspect of Leroi-Gourhan's thought: his systematic use of the unmarked concepts of equilibrium/disequilibrium (on which point, see Johnson 2011: 475, 484–7). For Leroi-Gourhan the palaeontologist, the idea of an increasing distortion in the relationship between the biological human and its technological systems presupposes a more 'natural' point of equilibrium, a prehistorical or historical stage at which the human body and human mind are more in balance with their externalised (humanised) environments. For Leroi-Gourhan the anthropologist, such an equilibrium equates at the level of the social with the individual's psychological integration into an ethnic unit of a tolerably 'human' scale. Placed in its historical context, the anxiety expressed in the second part of *Gesture and Speech* concerning the contemporary 'fate' of *Homo sapiens* derives from the state of accelerated technological development initiated by the Second World War and the postwar 'planetarisation' of Western civilisation – what Lévi-Strauss termed 'monoculture' (Lévi-Strauss 1992: 38/36–7). In his own, dramatised (italicised) references to our world situation *today*, Stiegler shadows this discourse of disequilibrium (distension, distortion, disproportion) with his own discourse of 'disorientation':

the *who* is dimensionally out of proportion and temporally out of step with the *what* – the *Gestell* – of contemporary technology (TT2, 73, 81/91, 99).

The questions that Leroi-Gourhan – and Stiegler reading Leroi-Gourhan – raise about 'the end of the human' are of course legitimate questions: it is difficult to dismiss or relativise the globally transformative effects of the technological revolutions of the second part of the twentieth century. However, it can also be said that Leroi-Gourhan's evaluation of the fate of *Homo sapiens* follows an entirely conventional history of thinking about technology, one which contrasts the ascendent trajectory of the technical tendency – the autonomous and infinite perfectibility of technology – with the descendent trajectory of the human, the catastrophic history of societies and civilisations. It is here in fact that Leroi-Gourhan can be said to be closest to the Rousseau of the *Second Discourse*.[11] It is here also that he seems close to the argument of contemporary evolutionary psychologists that there is a mismatch between the biological human, which reached its point of stabilisation during the Upper Palaeolithic period, and the social and technological environments of the modern world (see, for example, Pinker 2002: 219–22). If there is a criticism to be made here of Stiegler's reading of Leroi-Gourhan, it is that his generalised references to the 'speed' and 'acceleration' of contemporary technological civilisation too readily replicate these arguments without critical qualification. His evaluation, following Leroi-Gourhan, of the 'dis-ease' (*mal-être*) of contemporary existence, of 'a humanity that is essentially a latecomer [*retardataire*]' (TT2, 95/115), is consistent with the existential mode (and mood) of his analysis, but leaves open the whole question of the essential 'nature' of the human and the specific realities of modern technological systems.[12]

To conclude, viewed from a wider perspective, Stiegler's engagement with the anthropology of Leroi-Gourhan provides an interesting case study of the ongoing dialogue between philosophy and the human sciences in France. In 'Of Grammatology as a Positive Science', Derrida had proposed a mutually correcting or regulating relationship between grammatology and the 'facts' of scientific knowledge: 'a reflection must clearly be undertaken, within which the discovery of "positive" facts and the "deconstruction" of the history of metaphysics, in all its concepts, are subjected to a detailed and arduous process of mutual verification' (Derrida 1997: 83/124). Stiegler's dialogue with Leroi-Gourhan is a compelling example of the extension of this project to the question of technology, in which Stiegler's own deconstruction of the metaphysics of presence finds its scientific grounding in Leroi-Gourhan's materialist description of human origins. As has been seen, this dialogue enables Stiegler to develop a number of concepts which will become central to his philosophy of technology. At the same time, Stiegler's reading is critical of the residual essentialism of Leroi-Gourhan's anthropology, questioning his 'oppositions' between different types of prehistoric humanity. The conclusion of the analysis above was that this correction of anthropology is coherent within its own parameters, but that its critique of Leroi-Gourhan is not a definitive one. More interestingly, it revealed the possible divergences between philosophy and anthropology in terms of their different framings of the question of time and the nature of the symbolic – from this point of view, and following Derrida's remarks above, it could be said that anthropology might in its turn perform a corrective function in relation to philosophy. Finally, our analysis indicated a relative lack of critical perspective in Stiegler's replication of Leroi-Gourhan's diagnosis of the dystopic development of modern humanity. Stiegler's reading of some of the more speculative passages of *Gesture and Speech* allows him to develop his own narrative of 'disorientation', but it is not clear in the final analysis how far this diagnosis advances our understanding of the specificities of contemporary machine civilisation. Despite these reservations, the strength and the value of Stiegler's engagement with Leroi-Gourhan in *Technics and Time* is that it raises some challenging questions about the origins and ends of humanity, encouraging further critical reflection on the nature of technology and the nature of the human.

Notes

1. Richard Beardsworth notes this intellectual genealogy in an early appreciation of *Technics and Time, 1* (Beardsworth 1995: 90–1).
2. For Leroi-Gourhan's quasi-Bergsonian definition of 'tendency', see Leroi-Gourhan 1973: 336–7.
3. The genus name *Zinjanthropus* is no longer used in palaeoanthropology, and has been replaced by *Australopithecus* or *Paranthropus boisei*. Since Stiegler's reading of Leroi-Gourhan does not make this adjustment, for the purposes of consistency the following analysis will retain the older term.
4. Leroi-Gourhan uses the now outdated categories of Australanthropian,

Archanthropian, Palaeoanthropian and Neanthropian to designate the principal morphological stages of hominid evolution, the latter two stages relating to Neanderthal and *Homo sapiens* respectively.

5. The English translation of *La Technique et le temps* cannot capture the graphic and phonetic resonance of *silex* (flint) and *cortex*, which Stiegler visibly exploits in the original French text. It must be said that Stiegler's repeated use of this formulation (*silex/cortex*) can give the impression of a scene of technology restricted to lithic materials, whereas early hominid technical activity would obviously have involved a range of materials of differing degrees of perishability, all of which would have been capable of constituting an externalised 'memory' in the sense described by Stiegler. Stiegler's aphoristic formulation of the 'mirroring' of flint and mind should therefore be read as being metonymically implicit of this wider ecology of prehistoric technology.

6. For a useful discussion and contextualisation of this concept, see Nathan Schlanger's chapter in Audouze and Schlanger 2004.

7. Contemporary research in the field of language evolution takes seriously the idea of a 'protolanguage' and the existence of stages of development of prehistoric human language. See, for example, Knight, Studdert-Kennedy and Hurford 2000; Wray 2002; Christiansen and Kirby 2003.

8. 'Language can only have arisen all at once [*le langage n'a pu naître que tout d'un coup*]' (Lévi-Strauss 1978: 59/xlvii). Derrida will of course question this argument in *Of Grammatology* (Derrida 1997: 120–1/177).

9. On the distinction between linear (phonetic) and multidimensional (graphic, figurative) expression, see Leroi-Gourhan 1993: 195–6/270–2, vol. 1. Leroi-Gourhan's remarks on the symbolic and the figurative in *Gesture and Speech* need to be related to his more wide-ranging studies of prehistoric art and religion in other major works published during the same period, in particular Leroi-Gourhan 1964, 1965.

10. While Leroi-Gourhan's chronology for the appearance of anatomically modern *Homo sapiens* is now outdated (see Randall White's introduction to Leroi-Gourhan 1993: xxi), the discipline of palaeoanthropology has remained paradigmatically consistent in its recognition of a behavioural revolution occurring during the Mid-Upper Palaeolithic period (from approximately 35,000 years ago onwards), involving the emergence of elaborate ritual, aesthetic production and advanced projectile technology.

11. Leroi-Gourhan confesses this ideological affinity with Rousseau in his interviews with Claude-Henri Rocquet (Leroi-Gourhan 1982: 53).

12. In his interviews with Stiegler, Élie During questions what certain readers might perceive to be Stiegler's 'hyper-philosophical' approach to the question of technology, asking to what extent it is able or willing to take account of the material diversity of contemporary technologies (PA, 20–1, 24–5).

3

Of a Mythical Philosophical Anthropology: The Transcendental and the Empirical in *Technics and Time*[1]

Michael Lewis

Introduction

Bernard Stiegler's *Technics and Time, 1: The Fault of Epimetheus* is a reinvention of *philosophical anthropology*. The book's central thesis is that man never exists without technics, and this means that any *transcendental* account of man's emergence must implicate an *empirical* account of the emergence of technology. These two accounts together comprise a philosophical anthropology, but Stiegler shows that such an anthropology can only take the form of a *myth* – and it is in this that Stiegler's 'reinvention' consists.[2] Martin Heidegger has shown that in order to be individuated, the human subject must relate to itself in a way that is temporal because it joins the self as it presently is to the self as it will be in the future. Death is that event beyond which the self cannot experience the future, for any time that unfolds beyond that moment cannot be *my* future, and so death stakes out that stretch of time which defines my individual existence. In relating myself to death, I can relate only to myself, and thus I become most properly myself within this reflexive relation.

The subject's relation to itself is thus temporal, but Stiegler advances some way beyond Heidegger in showing that the human being is not by itself capable of achieving this 'transcendental subjectivity', which is 'transcendental' in the sense that it would be capable of constituting the objects of its own experience. Stiegler shows that man can relate to time only if he is already involved with 'technics' (*la technique*). 'Technics' is an obsolete English word that is used to translate a modern French term which encompasses techniques, technology, and the objects produced by these means: it thus includes the objects of pre-modern craft, pre-industrial and industrial techniques, and modern machine-powered technology.

These techniques and technical entities have a dynamic of their own, the objective expressions of which constitute a history (*Geschichte*). Thus, in Heidegger's terms, there is no temporality (*Zeitlichkeit*) without historicality (*Geschichtlichkeit*) (Heidegger 1962: ¶¶72–4; TDI, 160–1). This history of technology is, crucially, an *empirical* history, which means that it cannot be deduced *a priori* as a 'transcendental history' can. Stiegler shows that the transcendental can close upon itself only by encompassing the brute empiricity of the stone, as an oyster enclasps a speck of grit, and is thus able to consummate itself in the production of a pearl.[3]

In this way the technical object allows the human being to relate to time while simultaneously anchoring this relation within a history. This goes some way towards explaining why Stiegler describes his work as an '*archaeology* of reflexivity' (TT1, 140). Reflexivity has an archaeology because this reflection takes place only by way of the tool, and these tools are then preserved for the future archaeologist to discover. 'The analysis of the technological possibilities of the already-there [the historical-technical] peculiar to each epoch will, consequently, be that of the conditions of reflexivity – of mirroring – of a *who* in a *what*' (TT1, 237).

This is why we describe Stiegler as a philosophical anthropologist, because neither philosophy alone, if it is understood as transcendental, nor anthropology alone, if it is understood as empirical, can provide us with an adequate account of a humanity that is inextricable from technicality. At the same time, we must ask why Stiegler himself refuses this description and refers very little to its exemplars.

One of the keys to our interpretation is the notion that the narration of the origin of the human always takes place retrospectively, and that means from a particular perspective. The perspective Stiegler chooses is that of contemporary technology. A certain technological system is beginning to corrupt the individual's relation to its future and therefore needs to be addressed: the stone at our hearts is threatening to supplant the living organ altogether.

Transcendental anthropology

There are two traditional approaches to the origin of man: transcendental and empirical. Broadly speaking, and not without certain important nuances, Jean-Jacques Rousseau may be identified with the former and André Leroi-Gourhan with the latter.

In order to reach a proper understanding of the human it will be crucial to see how *neither* of these approaches can succeed because both fall short of the contemporaneity of man and tool.

Despite Lévi-Strauss's insisting that Rousseau is the father of *both* scientific and philosophical anthropology, Stiegler for the most part takes him to represent *philosophical* anthropology, which is to say a 'transcendental deduction' of the conditions for the possibility of man (TT1, 85). This deduction leads Rousseau towards a primitive humanity with qualities that must characterise every human being, because they precede the differentiations introduced into the human species by the 'technical' supplements to his nature, brought about by particular cultures and their history. This particularisation results from man's leaving his original habitat and dispersing into multiple geographical locations and climates. The first man possesses a naturalness that is only corrupted when he wanders away from his origin, into a differentiation which shatters his universality and adulterates the noble savage's original purity.

Thus the transcendental anthropology introduces a whole series of oppositions, beginning with nature and culture, such that it can hardly avoid thinking of pre-human animals and human beings as opposed to one another in the precise sense that each excludes the other and is defined by this negativity (TT1, 108). But, in a move that is crucial to Stiegler's own rethinking of the transcendental project, Rousseau undercuts this oppositional understanding: the original state of man that he speculatively posits in his *Essay on the Origins of Inequality* (1755), a primitive universality without cultural differentiation, is not stated to have existed *in fact*. Indeed this is why the origin cannot be ascertained by facts and requires a transcendental deduction (Rousseau 1984: 78). As Stiegler puts it: 'The essence (the origin), impossible to find in the facts (the fall), calls for [. . .] a transcendental recollection' (TT1, 108). This transcendental story will precisely amount to a 'necessary fiction' (TT1, 108). Stiegler will call it a 'myth', and the lesson he takes from the 'father of anthropology' is that the transcendental and the mythical *converge* when it comes to the question of man.

The transcendental myth proposes a stage of humanity free of all technics, and this state would comprise the 'origin' of man. But this is just the first stage in the origination of man, for he is not composed solely of this innocent purity; there is also a fall. The origin thus has two distinct moments, and it is this duplicity that Stiegler finds in both transcendental and empirical anthropology. It is a doubling of origin which he will associate with the origin's mythologisation. Stiegler's first substantial reference to the myth of Prometheus and Epimetheus arises in the context of an interpretation of Rousseau: the myth depicts a sequence that runs from a human animal that lacks to a human being whose lack is supplemented by technics. Myths are chronological stories and they describe an origin in narrative terms, and the myth which best expresses Stiegler's insight tells of Epimetheus's forgetful lack of foresight leaving man bereft of qualities and his brother Prometheus's theft of fire from the gods (TT1, 113).

Empirical anthropology

Nevertheless, Rousseau himself does not deploy this myth, and his account falls prey to an illusion to which all transcendental anthropology must succumb. Transcendental anthropology determines the origin of man *retrospectively,* and this approach can be decisively criticised by approaching this origin from the other direction, *prospectively,* from before it has happened. Rousseau's mistake is to begin from man as he now stands, and to treat him as if he had always stood and walked upright, but without using his newly liberated hands for the manipulation of tools (TT1, 113; Rousseau 1984: 81–2). For Stiegler, 'Rousseau may well decide to ignore the facts; he may not, however, totally contradict them' (TT1, 112).

Rousseau does indeed refer to the factual discipline of 'comparative anatomy' but dismisses it because of its relatively inchoate state (1984: 81). The work of André Leroi-Gourhan, who uncovered facts that directly contradict Rousseau, means that this inchoateness is no longer an issue. Leroi-Gourhan has demonstrated that the upright stance of the primitive human being freed the hand from the tasks of walking and fighting, reassigning its function to the grasping of tools. The empirical anthropologist thus proves that the emergence of man begins when the quadruped becomes the biped, and this may be shown empirically to be strictly contemporaneous with the emergence of technics. He thus overturns Rousseau's hypothesis of an upright man *without* technics: 'the upright position has a meaning and consequences that are incompatible with Rousseau's account of the origin of man' (TT1, 113). Empirical palaeoanthropology reveals that there is no human being without technicality. By virtue of this insight, 'palaeontology will profoundly affect the anthropological *a priori*, governing at the most profound level the most authentically philosophical questioning' (TT1, 132). It does this by insisting that our definition of man must rule out the possibility of a man *without* technics. It is precisely this thesis that generates the very structure of Stiegler's work.

The problem for Leroi-Gourhan is that humanity cannot be understood as a zoological species in any obvious way (Leroi-Gourhan 1989: 48–50). Man can be theoretically unified only by reference to the use of tools. In him, the evolution of animal life continues, but at a different rhythm to that of genetic drift. This is because technical objects display their own, non-zoological tendencies. Thus, in man, 'the evolution of life continues by means *other* than life' (TT1, 135). This is another reason why it is problematic to speak of the human being as an animal species: man constitutes a continuation of life's evolution, but by radically new means opposed to the old – exteriorised technics, epiphylogenetic memory, the novel inheritance that transmits individual experience (epi-) to subsequent generations of the species (-phylo-) and thus initiates history in the strict sense. This idea allows Stiegler to maintain that the relation between man and animal is both an opposition and a continuation, for this form of memory 'must not be understood as a rupture with nature but rather as a new organisation of life – life organising the inorganic and organising itself therein by that very fact' (TT1, 163).[4]

Man thus lacks a specific difference and identity *until* he is reflected in the technical artefacts that he produces (and which thereby simultaneously produce *him*). In a way that is avowedly indebted to Lacan, Stiegler understands man to be unified by an original 'absence of propriety [or properness]' (TT1, 133). This can be remedied only if it is supplemented by technics: ' "human nature" consists only in its technicity, in its denaturalisation' (TT1, 148; see also TT1, 157, 216). Man needs external technical objects to act as mirrors that reflect him and thus allow him to acquire a reflexive identity that he did not previously possess. The tool is in this sense a 'proto-mirror', and this period of man's evolution amounts to a 'proto-mirror stage', which for Stiegler is just as much a part of phylogeny as it is of ontogeny (TT1, 157).

Stiegler's text devotes itself to the task of remaining true to the thesis of the contemporaneous origination of man and tool, and to showing why an account of the origin of man must nevertheless slip into a mythological chronologisation of man and tool whenever it attempts such a task. Even Leroi-Gourhan is at a certain point snagged on this mythopoietic machine, along with his opponent, Rousseau. Rousseau postulates the existence of a non-technical man at the beginning of his story, while Leroi-Gourhan does the same at the end (TT1, 151; Leroi-Gourhan 1989: 92–3). So, in both empirical and transcendental anthropology, the origin of man is split into two stages.

From anthropology to philosophy

Stiegler's book is constituted by the effort of resisting this apparently irresistible duplication. Were a theory of man to achieve this, it would have moved definitively beyond an anthropologistic understanding and become 'philosophical'. Stiegler expresses the movement as follows, in reference to Heidegger's 'analytic of existence': 'is not the consideration of *tekhnē*, as the originary horizon of any access of the being that we ourselves are to itself, the very possibility of disanthropologising the temporal, existential analytic?' (TT1, 262). Despite Heidegger's own attempt to produce a non-anthropological understanding of man, he failed in precisely the way we have seen Rousseau and Leroi-Gourhan fall short, by ultimately failing to acknowledge the co-originarity of human and the tool, temporality and historicality. In anxiety, I relate solely to my own death, and the world seems to slip away. Thus, in my most authentic state, the ready-to-hand shows itself as inessential, and with its vanishing, history is likewise eclipsed (Heidegger 1998: 88).

So, for Stiegler, 'philosophy' designates Heidegger's approach to the question of man but perfected in light of a thesis drawn from anthropology (Leroi-Gourhan), one which immediately compels the latter to exceed its own (zoological) limitations. A philosophical understanding of the human being takes it to be something resolutely non-anthropological and non-animalistic, while nevertheless remaining within a history of animal life, albeit in the form of a technical exteriorisation of that life. Thus, in Stiegler, we find a philosophical anthropology that becomes philosophical precisely in realising that anthropology must of necessity exceed itself once it recognises the essential function of technics in relation to life. Stiegler clarifies the transition from anthropology to philosophy as follows: 'any residual hint of the anthropological is abandoned through the fact that technology becomes properly speaking a thanatology' (TT1, 187). This means that technology is here understood as making possible a relation to time and first of all to the future, and the future when it is understood in relation to individuation is death (*thanatos*). This is counterposed to a thinking of anticipation that would understand it as a given quality of the putatively zoological species, '*homo sapiens*'. But with the intervention of a technical relation to *death*, the last preserve of non-technicality is irremediably lost.

The mythopoietic machine

The first part of *Technics and Time, 1* demonstrates the contemporaneous invention of man and tool, and the way Simondon, Rousseau and Leroi-Gourhan became afflicted with double vision when confronted with this fact. The next stage in Stiegler's itinerary is to reveal the mechanism which makes this not merely a fault of subjective perception, but an inevitable diffraction. Philosophers cannot but fall back into anthropology, and Stiegler sees the origin of man as a machine for generating myths.

In the second part of the book, 'The Fault of Epimetheus', Stiegler demonstrates the necessity of supplementing Heidegger's existential analytic with the myth of Epimetheus and Prometheus. The supplementation of Heidegger's theory is intended to show that the peculiar evolution of man produces a 'transcendental illusion' or mirage which makes it seem as if man could one day have been without technics, or would one day reach such a point. Thus *Technics and Time* tries both to remain true to the concomitance of man and tool, and to do so precisely by explaining why philosophical and anthropological accounts always fall away from this in a process of 'dephasing', which, properly understood, produces a myth of origin.[5] To achieve this, Stiegler must establish the correct relation between the *empirical* facts established by anthropology, and the *transcendental* approach taken by philosophy, and finally, to ascertain the connection between this relation and the mythical chronology to which it gives rise.[6]

Man and technics, transcendental and empirical

To progress we first need to establish how the co-originarity of man and technics amounts to a mutual contamination of the empirical and the transcendental.

If one were to consider Stiegler's work merely as a perfecting of Heidegger's existential analytic that situated the techno-historical entity at the heart of time, then 'considered *from this perspective*, epiphylogenesis is a transcendental concept' (TT1, 243 [emphasis added]). But this would ignore the most important philosophical consequence of rendering the auto-affective relation of temporality dependent upon technology, for 'this concept undermines itself at one and the same time, *suspending the entire credibility of the empirico-transcendental divide*' (TT1, 243 [emphasis added]).

This means that the perpetual failure to fully uphold the thesis of the contemporaneous invention of man and tool is due to a seemingly inextinguishable wish to restore purity to the opposition between the transcendental and the empirical since it involves preserving the transcendental *subject* from any empiricity and empirical historicity. In the modern age at least, this amounts to a fall into anthropologism since this transcendental subject is understood to be inextricable from the human being. Consequently, the falling away from the necessary duplicity of man and technics is a return to *metaphysics*, since even a pure positivism of facts is still metaphysical in so far as it posits the metaphysical thesis that the world is ultimately comprised of atomic facts, which involves a radical division between the essence (which is posited as nonexistent) and existence (factuality).

Leroi-Gourhan's palaeoanthropology claims to speak of the origin of humanity 'from outside the snare of metaphysics' (TT1, 84). Importantly, this means that it will try to place itself beyond both transcendental anthropology and positivist empiricism. Passing beyond metaphysics it seems that one is already standing astride the transcendental-empirical divide. Hence one will remain true to the insight of Leroi-Gourhan only by insisting upon a mutual *contamination* of the transcendental and the empirical (TT1, 84).

Let us first of all see how the transcendental is infected with empirical factuality (a), and then examine the idea that empirical factuality must also be infected with transcendentality (b). The second task is the more difficult because Stiegler is frequently understood as appealing to 'facts' pure and simple, which is what we must avoid at all costs if we are to remain consistent with his logic.[7]

(a) How the transcendental is infected with empiricity

Stiegler examines the historians of technology (Bertrand Gille, Richard Lefèbvre des Noëttes (TT1, 30), Gilbert Simondon) as well as its prehistorians (Leroi-Gourhan) in order to show that technics has a dynamics of its own which can be ascertained only *empirically*. Stiegler's transition from the technologists to the anthropologists is intended to demonstrate that the very autonomy of this external memory *defines* man. Subjectivity cannot be formed without the empirically historical technical object. The subject's temporal self-relation conditions the possibility of all experience and may therefore be deemed 'transcendental', and because the relation will not be completed if it does not enclose a historicity that is empirically conditioned, the transcendental will never have been purely transcendental; its very constitution *depends* upon the empirical.

(b) How the empirical is infected by transcendentality

In the context of the second task, Stiegler understands the relationship between the empirical and the transcendental in terms of the relation between facts and their interpretation: 'facts [. . .] *are only given against the background of possibilities of interpretation* that are not themselves of the order of facts' (TT1, 99). When it comes to origins, facts become intelligible only as part of an interpretation, made in hindsight from a particular point of view within the present moment. This raises the difficult question of how to cope with the potentially distorting effects of hindsight, and the risk of mistaking a (retrospective) transcendental condition for a (prospective) factual chronological one, if this error can be avoided at all.

On plasticity

What happens to the fact when it is considered within the context of a retrospective interpretation? First of all, why does it not remain a mere fact? To answer this, we need to specify the empiricity in question: it is the history of technics. Stiegler has already shown that technical evolution is independent of zoological evolution, but this distinction is made partly in order to demonstrate the particular way in which the two evolutionary tendencies *converge* to form the human being. This means that the zoological character of man must be *conducive* to such a relation and indeed partake of a specular exchange with the technical object: in particular, this means that the animal human being must have a particular kind of skeleton and brain.

Here it is useful to recall a remark that Stiegler makes almost in passing, regarding the 'plasticity' of the human brain: 'the appearance of these tools [. . .] supposes a singular epigenetic plasticity of the cerebral structure' (TT1, 176–7),[8] and to place this alongside his remarks on the non-specialised character of the human being's organs and members (TT1, 79, 155–60). However, this does not mean that Stiegler is adopting the zoological explanation which he elsewhere criticises, which states that the adoption of the technical supplement was a necessary consequence of the fact that the front paws and teeth of the pre-hominid had become useless as weapons when it assumed the bipedal stance (TT1, 150). Such an understanding of technics is zoological because it understands the technical as the solution to a biological problem.

So how are we to differentiate Stiegler's solution from this straightforwardly zoological account? For Stiegler, the empirical tool fills in a gap in the animal's self-relation, and in the present context we might understand this as an evolved lack of specialisation and rigidity, an indeterminacy of function, the empirical fact that the human animal developed these features in the course of its zoological evolution. The facts involved here are those which *allow* the tool to become part of the constitution of the transcendental subject.

We have already spoken of the way in which the transcendental subject is affected by this supplement, but what happens to the *tool* when it is understood as serving this function?

On taking tools for souvenirs

The empirical facts of the history of technics are selectively interpreted by Stiegler in relation to a certain end, and that is the role that the tool plays in the construction of the transcendental, and as a result only those facts which *can* be so interpreted are taken into account. Hence Leroi-Gourhan and Stiegler interpret the tool not as a means to an end – which it evidently also is – but as a type of *memory*, an enduring trace of the past: 'a tool is, before anything else, memory' (TT1, 254).

This is to take the tool in a way that is independent of the intention of the inventor with respect to the specificity of the tool and the particular use that the tool has, whether it be killing, cutting, scraping, boring, or even writing. It is instead to understand the tool simply as something that was made in the past to be used in the future (temporality), and something which stands as a record of an earlier stage of technical development, a trace that can be inherited by later generations (historicity): in this way, the tool, understood as an external memory, constitutes and knits together the temporality and the historicality in which the human being develops.

From the point of view of transcendental subjectivity, 'the invention of the human' (Stiegler's title for the first part of *Technics and Time, 1*), the tool is of interest only as a temporal and historical object, an epiphylogenetic memory. The originality of Stiegler's theoretical gesture is expressed in the very unfamiliarity of taking a tool as a memory, a truly peculiar idea.

Here as so often, Stiegler follows Heidegger only to exceed him: the instrumental–anthropological interpretation of the tool may be *correct*, but it is not *true* (Heidegger 1993: 313). The technical object need not be viewed solely in terms of how man uses it, but also in terms of what it reveals, and indeed what it reveals about man and the constitution of a reflexive subject. What it reveals is that, in man, life's evolution is prolonged in a new, epiphylogenetic form.

So the empirical facts of the history of technics would be transformed by being incorporated into the genesis of transcendental subjectivity, and this transformation is expressed in the fact that Stiegler interprets them as memories, and thus in terms of time. How crucial this move and this effect are can be measured by the very title of Stiegler's series: technics and time.

Now we are in a position to understand Stiegler's assertion that Leroi-Gourhan avoids treating the facts he reveals in a positivistic manner. This is crucial to our reading, and our hypothesis is precisely that Leroi-Gourhan adopts one very particular *perspective* on the facts that we have just outlined, the perspective of contemporary technology. Stiegler adopts the same perspective, and he does indeed speak of these facts, whilst also adducing some of his own regarding neural plasticity. Both interpret the prehistoric evolution from the quadrupedal to the upright stance from the perspective of the ultimate assumption of the tool by the hand. Thus they can read the human as beginning with the feet, but only because this interpretation is retrospective and adopts the perspective of a particular *interpretation*. There are indeed facts here, but they cannot be knotted together to form an explanation unless one takes up a particular point of view that allows one to order these facts into something like a story. It is the necessarily retrospective character of interpretation – the transcendentality of the empirical – that Stiegler tries to capture by speaking of the origin in terms of *myth*.[9]

'Lack' as mythical

This might allow us to approach the question of why Stiegler does not ally his own approach with those discourses which seem so close to his own, that are inspired by the natural sciences and anthropology and which understand man as unable to survive without an extended period of technical or prosthetic support: 'philosophical anthropology'.[10]

For Stiegler, the problem with these discourses is the way in which they understand the relation between the factual, natural lack and its technical supplement: they do not have an adequate understanding of the transcendentalisation of the empirical that we have outlined in terms of facts and their interpretation. They fail to see that their discourse can only be mythical. The (human) animal can only be described as 'lacking' in hindsight, from the point of view of a human being which has already been supplemented by technics: transcendentally, and not purely chronologically, which is to say from the point of view of the beginning. Stiegler's implicit response to the philosophical anthropologists is to suggest that the very form of their discourse risks 'naturalising the lack'.

Stiegler calls these two moments of lack and technical supplementation by the mythical names of Epimetheus and Prometheus. He does this in order to indicate, first, that the sequence is not *really* chronological (man and technics arise contemporaneously), and, second, that it is nevertheless necessary to narrate the sequence in this way. This means that the empirico-transcendental contamination that defines the human itself automatically generates a mythical retelling of its own origin. This automaton dephases the onefold of man and technics into a diffracted and animated mirage which always sees in itself a movement from lack to its technical supplementation.

Stiegler explicitly avows that the co-invention of human and tool produces 'the illusion of succession' (TT1, 142). Rousseau's enduring merit, for Stiegler, is to have noticed the necessity of this fiction. It is impossible to decide which comes first, subjectivity's need of being supplemented by empirical technics, or the tool's assumption of a transcendental function: it is an 'aporia', a moment of undecidability in which neither of two opposed routes is navigable: 'the aporia always ends up hardening into a mythology that opposes two moments [. . .] This is an excellent archetype of the discourse of philosophy on technics, relating through a fiction, if not by a myth, how the man of pure nature is replaced by the man of the fall, of technics' (TT1, 101).

Conclusion: Of an anthropology that must be philosophical

Ultimately we may say that what makes Stiegler's work so unusual is the curious coincidence of a retrospective and a prospective vision of man – a transcendental philosophical and an empirical anthropological perspective. This is why it is quite consistent to describe what Stiegler is doing as a philosophical anthropology.

The mythical clothing of an anthropological account is the *formal admission* that the story is told retrospectively, with a backwards glance towards the evolution in which the human emerges from the animal, an account which interprets the facts of neurobiology, anatomy and zoology – the animal facts from a time before 'man'. It is thus a hindsight, which impacts upon the *pro*spective standpoint that one should adopt.

This explains why Stiegler can speak of the relation between man and animal as *both* continual *and* oppositional, for the completely novel epiphylogenetic memory carries forward a movement of life that traverses all animality. We can speak of a radical leap only by viewing the history of animals from the point of view of technics, but this remains a continuist reading of this very movement. Perhaps what allows it to be so is the Derridean nature of Stiegler's understanding of life, which takes it to be homogeneous, at least up to a certain point and on a certain level, as *différance*, a medium in which recording by means of a trace or 'memory' can take place. The first stage of deconstruction is to efface the single line of opposition that metaphysics posits between man and animal, to render the field of all animals, including the human animal, homogeneous; but this is done only for the sake of a second stage in which one opens up a diverse typography within this realm (Derrida 2004: 72–3; Derrida 2008: 47–8; Derrida 2009: 15–16).

The difference between man and animal is rendered *undecidable*, but only in order to open up the possibility of multiple *decisions* which will result from a multiplicity of different perspectives taken upon the newly homogenised realm as a whole. Stiegler's decision is to examine this terrain from the standpoint of modern technology, and consequently he deploys a history of the way in which the inscribed mark, understood as 'orthothetic' or increasingly accurate memory, develops from the simplest forms of life to the most complex, and for this reason he is forced to *restore* a kind of opposition between man and the rest of the animals.

Thus Stiegler's work is a reinvention of philosophical anthropology; but this discourse can only be enunciated in a mythical form, which expresses the mutual entwinement of the retrospective and the prospective, the transcendental and the empirical, that follows from the originally (palaeo)anthropological thesis that man and technics are born as one.

Notes

1. Earlier versions of the ideas found in this essay were presented before audiences at the University of Cambridge (Anthropology) and the University of the West of England (Philosophy) on March 15th and May 2nd 2012, respectively.
2. One of the earliest accounts of Stiegler's work adopts the same perspective and remains one of the best (Beardsworth 1995: 2–4).
3. Tracy Colony provides an excellent account of the way in which Stiegler accompanies Heidegger's use of concepts only in order to take them beyond him at a certain point (Colony 2010: 120–1), before going on to suggest a Heideggerian response that demonstrates the inherence of a certain facticity in Heidegger's own account of time (2010: 122–4).
4. This is why Stiegler goes on to say that '*Nature must be understood*

differently', beyond its opposition to technics (TT1, 163). He immediately specifies, however, that this slackening of the absoluteness
of the opposition between nature and culture does not homogenise
the entire animal realm and eradicate differences. In this, Stiegler is
a pupil of Derrida. But unlike Derrida, Stiegler is not content simply
to multiply the differences notionally and to establish the possibility
of a more complex typology; rather, he makes a *decision* and adopts
just one of the many possible perspectives on the animal realm, one
in which man and animal are indeed opposed to one another in a
certain way. Derrida himself comes very close to explicitly allowing
this as a possibility within a *certain* taxonomy (Derrida 2004: 72–3).
Our argument will be that Stiegler makes a decision that leads to just
such a taxonomy.

    The *decisive* moment of Stiegler's relation to Derrida seems not to
be taken into account by Ben Roberts's critique of Stiegler's reading
of Derrida (Roberts 2005). The exact relationship between Derrida
and Stiegler on this point will be addressed in a future work.

5. I refer to Gilbert Simondon's notion of an entity 'falling out of
   phase with itself' (*déphaser*) (Simondon 1992: 300). The difference
   between Stiegler's use of the term and Simondon's is that Simondon
   refers to a real or a logical succession, while for Stiegler the appearance of succession is purely mythical.

6. For a treatise on anthropology, philosophy and myth, Stiegler's
   references to Claude Lévi-Strauss are surprisingly scant. In *Technics
   and Time, 1* they number only two or three (TT1, 93, 101). Stiegler
   seems to suggest that the fault of Lévi-Strauss is precisely to repeat
   old father Rousseau's mistake of installing a rigorous distinction
   between the empirical and the transcendental, mirrored in a strict
   separation between nature and culture that results from the incest
   taboo and the symbolic law which develops on this ground, incorporating all kinds of non-instinctual prohibitions. This leads Lévi-
   Strauss to postulate a *simple* origin of man and therefore a *human
   nature* which would be a universal invariant of every human culture.
   It is this unity of man that Leroi-Gourhan puts in question, which
   is why his work is more interesting to Stiegler than Lévi-Strauss's
   (TT1, 93). But is Lévi-Strauss not himself interested in the mythical-
   ity of any attempt to suggest that the incest taboo could have been
   instituted all at once, at a moment that can be chronologically
   located? Even Stiegler gestures towards such a reading (TT1, 101).

7. Therefore our explanation of the transcendental–empirical contamination will take us some way towards refuting Geoffrey

Bennington's critique of a certain 'positivism' and a confusion of the 'quasi-transcendental' and the empirical (or 'transcendental contraband') in Stiegler's text (Bennington 1996: 190, 195–6), which David Wills describes as the 'palaeontological and the ontological' (Wills 2006: 240, see also 260–1 n. 12). Even Beardsworth, among the most sympathetic of Stiegler's commentators, comes close to this form of objection in suggesting that Stiegler restricts the notion of originary technics (the 'quasi-transcendental') to just one empirically specifiable *kind* of technical objects, those involved in the hominisation process (Beardsworth 1998a: 81).

8. He also speaks of a 'double plasticity' of *both* cortex and flint (TT1, 142, see also 135). Stiegler may well be alluding to Catherine Malabou's work in this context (see TT1, xi).

9. This is where the present essay differs from Colony's critique of Stiegler, which accuses him of an anthropocentrism that opposes technical man and non-technical animal, ignoring Derrida's insight into the dead elements to be found in *all* life. For Colony, this results from Stiegler's conflating the non-living with the technical in the *narrow* sense of human techniques and artefacts (Colony 2011: 85, see also 75, 84). A similar thought, according to which Stiegler restores a radical opposition between the human and the other animals, may be found in Roberts (2005), in a part of the text which bases itself largely upon Beardsworth's work (1998a: 81), which is itself derivative, according to Roberts (Roberts 2005: n.18). But this is precisely what Stiegler avowedly wishes to do, and he is allowed to do it by virtue of his decision regarding the perspective from which to view and taxonomise the animal realm (contemporary technology), and the mythical character of his discourse here. Colony himself says as much, asserting – in a *critical* way – the mythical character of Stiegler's thought at this point (2011: 88, 86).

10. In Stiegler we find only one, indirect reference to Arnold Gehlen, the greatest of the twentieth-century German philosophical anthropologists (TT1, 11).

4

Technics and Cerebrality

Ian James

In the two volumes of *De la misère symbolique* (*Symbolic Misery*, 2004; 2005) Bernard Stiegler begins to sketch out one of the key concepts which has inflected the development of his thinking in recent years, that of 'une organologie générale', or general organology (SM1, 14fr). The organs that are to be thought in the light of this concept are not simply the physiological constituent parts of the human body. Indeed organology could be said to be 'general' here in so far as it interrogates not only physiological organs, but also 'technical' organs as well as forms of social *organ-* isation. As a concept general organology is a framework which allows the interrelation of physiological and technical objects to be understood and circumscribed. This, in turn, allows the techni- cal or technological ordering of specific (historical) societies to be determined.[1] Following on from Gilles Deleuze's thinking of the 'control society' (Deleuze 1995), the interest of organology lies, Stiegler argues, in its potential to allow us to 'understand the hist- orical tendencies which have led to the specificity of our present time' (SM1, 14fr). Later in the first volume of *De la misère sym- bolique* he specifies further that what we need to understand is the 'general *organological horizon* with, for and against which it is necessary to struggle, in the name of what concerns *us*, and, in so far as we desire to remain an *us*, to remain individual singularities' (SM1, 101fr). Organology, then, takes as its object the historically contingent structures in which physiological sense organs are con- joined and co-articulated with artificial organs or what one might term the generalised field of 'technical prosthetics'. This interroga- tion, together with the understanding that might result from it, is resolutely in the service of a philosophical–political analysis of contemporary culture and, more specifically, it is in the service of a politics of individuation, that is to say, a politics of the way in which individuals singularise and differentiate themselves within any given social collective. Stiegler wishes to challenge the ways in which we are collectively conditioned by mass media and marketing technologies in particular and he aims to 'open the prospect of a *total organological revolution*' (SM2, 101fr [§22]).

One of the most important innovations that results from Stiegler's reworking of Heidegger and Husserl (via readings of, among others, Leroi-Gourhan) in the *Technics and Time* series, is the way in which it allows him to develop a historical perspective on different epochs of technological systems, modes of inscription, or regimes of writing. In Derridean terms, and in a divergence from Derrida's thinking, Stiegler is able to account for and describe different epochs of the ordering and structuring of *différance* or (arche-)writing (James 2010). His analyses of 'the technological rooting of all relation to time' (TT1, 135/146), which are concerned with a description of the fundamental (ontological/phenomenological) structures of world disclosure, shared existence and historicity, increasingly turn towards the philosophical–cultural analysis of our contemporary and emergent epoch of writing. It is in this context that the recently developed concept of organology occupies such an important place in Stiegler's thought. Yet the question of human bodily organs, their evolutionary development, their specificity and their relation to technical prosthetics is central to his thinking from the very first volume of *Technics and Time*. Before 'organology' becomes a key concept in Stiegler's armoury for his political and cultural struggle,[2] the interrogation of organs lies at the very centre of the initial philosophical work, which provides the orientation of his thinking as a whole.

The hand and not the brain

This is most evident in the sustained engagement with André Leroi-Gourhan which takes place in the first volume of *Technics and Time*. Drawing on Leroi-Gourhan's work, Stiegler gives a philosophical account of the advent of the human in the evolutionary development of early hominids. In this account the development of the upright stance, the manipulation of tools with hands, and the expansion of the cerebral cortex are all closely interlinked. Tool use permits the sedimentation and conservation of an impersonal or collective memory, that is to say, tools are always the bearers of memory traces. These are traces of past worldly and purposive interactions that are conserved in the utility of the tool itself. According to this account the general field of technical prosthetics is one in which meaning traces are impersonally and collectively retained in inorganic materials, thus opening up a distinctively human, symbolic and temporal relation to the world (via what Stiegler comes to call 'epiphylogenesis'– see James 2010).

What is striking about Stiegler's reading of Leroi-Gourhan's bio-anthropological understanding of hominisation is the way in which he focuses on the account given of the development of the upright stance. There are indications that an expansion of the cerebral cortex may have accompanied the adoption of the upright stance in early hominids, but this, for Leroi-Gourhan and for Stiegler, in no way affirms a primacy of purely cerebral or cognitive capacity when it comes to defining the specificity of the human. The key organs that are at stake here are *first* those of the feet and then the hands but not the brain, as Stiegler makes clear, loosely citing Leroi-Gourhan: 'The beginning of the human is here, with the feet and not the brain'. The adoption of the upright posture frees the hands from the functions of supporting movement and orientates them more exclusively towards the advanced prehensile capacities associated with tool use: 'Because, *using hands*, and no longer having paws, is to manipulate – and what hands manipulate are tools, instruments. The hand becomes a hand in so far as it gives access to ways of doing [*à l'art*], to artifice and to *tekhnē*' (TT1, 113/124). Further on in his discussion this leads Stiegler to conclude unambiguously that the human: 'did not begin with the brain but with the feet [. . .] The brain evidently plays a role – but it is not the principal determination' (TT1, 145/155).

It is clear, then, that Stiegler's thinking in relation to bodily organs in the first volume of *Technics and Time* squarely lays the emphasis on the fundamental role played by the worldly, technical and purposive interactions that the upright stance and the freeing-up of the hands for more developed tool use permits. The cognitive capacities of the brain and the primary role it might be said to play in the constitution of human specificity is displaced in favour of the hand, its manipulation of tools, and the horizons of temporality, meaning retention and symbolic activity that tool manipulation makes possible. This is entirely consistent with the importance of the role played by Heidegger's thought in the first volume of *Technics and Time* and, more specifically, by the Heideggerian account of the ecstatico-temporal horizon of Dasein.

For both Heidegger and Stiegler, the privileging of the brain and its cognitive capacities or operations is a key element of a metaphysical conception of the human. Such a conception accords a privileged role to a calculative, mathematising and instrumental relation to the world, views the brain and its cognitive activity as the site and ground of a primarily calculative mode of subjectivity, and poses subjectivity itself as the ground and foundation of the being of beings.[3] By understanding Dasein as a manner of being which opens up in the temporalising projections of being-in-the-world the primacy of the brain is overturned in both Stiegler and Heidegger.

Indeed, as Derrida has clearly shown in one of his short essays on Heidegger, the hand plays a fundamental role in the Heideggerian account of being and of Dasein.[4] Derrida's deconstructive reading of the role played by the hand in Heidegger's thinking sheds important light on Stiegler's formulations in *Technics and Time, 1* and on what is at stake in the privileging of the tool-wielding hand over the brain. Two key points in relation to Derrida's account of Heidegger's hand are worth underlining in this context. First, for Heidegger as for Stiegler, one cannot engage philosophically with the significance of the hand without engaging also with the question of the human in general and, more specifically, with the question of technicity or tool use. As Derrida puts it: 'at issue is an opposition that is posed very classically, very dogmatically and metaphysically, between a man's hand and an ape's hand' (Derrida 2007: 35/42). He also adds: 'One cannot talk about the hand without talking about technology [*la technique*]' (Derrida 2007: 36/43). As soon as one interrogates the hand one places oneself within the legacy of anthropological and metaphysical determinations of the human (as opposed to the animal in general) and one also necessarily raises the question of specifically human tool use. It is clear that Derrida perceives Heidegger's relation to this metaphysical legacy of the determination of the human (hand) in opposition to the animal (paw) as being un-thought and that Heidegger's thinking of Dasein therefore remains very much within the orbit of the anthropological conceptions of human being that he seeks to overturn or transcend. For his part Stiegler insists that Leroi-Gourhan's bio-anthropological perspective depends on an essentially non-anthropocentric concept of the human which will allow him to think the question of technicity and human specificity outside of any traditional anthropologism (TT1, 137/148).[5]

Secondly, it is also clear that Heidegger is thinking the hand, not simply in its ontic determination, as one organ amongst others that could be examined or treated as an object beneath a scientific or medical gaze, but rather in terms of its fundamentally onto-logical status. This is because the human hand is, at a primordial level, defined in its specificity by its relation to the human capacity for thought, as Derrida underlines citing Heidegger himself: 'To think is a handiwork [*penser, c'est un travail de la main*]' (Derrida 2007: 35/45). So if the hand always articulates a relation to tech-nology or technicity in the Heideggerian text, then this technicity is always that of thought itself and therefore necessarily also that of language or speech (as the medium of thought). Again, Derrida underlines this very explicitly: 'the hand of man is thought on the basis of thinking [*est pensée depuis la pensée*], but thinking is thought on the basis of speech or language' (Derrida 2007: 41/49; 2008: 41). This in turn means that the hand cannot be an organ like any other. Derrida, for instance, draws attention to the fact that Heidegger speaks always of 'the hand' in the singular and not of 'hands' in the plural. So for Heidegger, 'the essential co-belonging of hand and speech, man's essential distinction' (Derrida 2007:47/54) confers on this organ an ontological status, in so far as it plays a fundamentally disclosive role. The essential co-belonging of hand and speech is the condition of possibility of thought and this makes Dasein a mode of being which can reflect on the question of its own being and therefore differentiates its mode of presence or way of being from that of the objects it encounters: 'Dasein is neither *vorhanden* nor *zuhanden*. Its mode of presence is other, but it must indeed have a hand in order to relate itself to other modes of presence' (Derrida 2007: 44/51).

The fact that the hand in Heidegger is fundamentally related to the technicity of language and thought on the one hand and that it has an ontological and not ontic status on the other is decisive for Stiegler's repetition of, and deviation from, Heideggerian thought in *Technics and Time, 1*. It might immediately be noted, for instance, that the essential co-belonging of hand, speech and thought in Heidegger can be related directly to his theses on modern technology, poiesis and on thought as a poetic 'saying' of being. Heidegger's complex thinking of modern technology as 'enframing' cannot be rehearsed properly here. However, 'enfram-ing', *das Gestell* in German, can most straightforwardly be under-stood as the mode by which modernity encounters or reveals the world as a site of calculation and exploitation. Heidegger makes a distinction between the modern technological revealing of the world as a 'challenging-forth' and the more primary or fundamental, 'poietical' revealing of the world understood as a 'bringing-forth' (Heidegger 1993: 307–42). The 'challenging-forth' of the world discloses nature as a site from which resources can be extracted, stored and then used or exploited (nature is treated in Heidegger's terms as a 'standing reserve'). The more primordial 'poietical' 'bringing-forth' of the world is understood as a creative unfolding or coming to presence; it is a creation or production of Being which has a care for the event or giving of Being itself and does not subordinate it to a calculative or exploitative logic.[6] So, 'bringing-forth' as a more primordial, poietical revealing of Being is aligned in very positive terms with art and the artwork (1993: 179) and is opposed to the (complex but largely negatively characterised) challenging-forth of modern technology. Readers of 'The Question Concerning Technology' and the 'Origin of the Work of Art' will know that Heidegger also aligns this more primordial revealing of being with '*tekhnē*', the ancient Greek term which refers to art but also to skill, craft or technique. That is to say he seeks to locate a more originary instance in which poiesis, art and what one might call technics, technicity or tool use are not opposed to each other. The word *tekhnē* reminds us that the opposition or division between the poietic as a bringing-forth and the technological as a challenging-forth was not always as entrenched as it is in the modern era.

The upshot of this is that when Heidegger aligns the hand with *tekhnē* he is also, *and decisively*, aligning the hand with art and with the more primordial and poietical revealing of being. The technicity of the hand is therefore concerned with 'poiesis', with the preservation of, and care for, the 'dignity' of the event of Being and not at all with the technological enframing or revealing of being. This, of course, explains why Heidegger's thinking and philosophical language themselves are aligned with poiesis, with poetry as designated by the German term 'Dichtung' and with a poetical saying of being. In so far as the hand in Heidegger is at once distinctive of the human, aligned with the technical and also fundamentally disclosive of being, it is first and foremost the hand of the artist, artisan or craftsman, and even more specifically it is a hand that holds a pen, quill or other writing prosthesis used by the poet/thinker. Only in a more secondary and derived mode is it a hand which wields the tools or instruments of modern technology.

It is here that the contrast with Stiegler becomes clearest and that the significance of Stiegler's difference from Heidegger can be most clearly highlighted. Once again it is the influence of Leroi-Gourhan that inflects the specificity of the Stieglerian thinking of the hand. In the development of the upright stance, tool use, and the concomitant expansion of the human cerebral cortex the relation between the human body and *all* technologies or modes of technicity is fundamental. Here no differentiation is made between the creative production of poiesis/*tekhnē* and the violent exploitation of modern technology. Stiegler cites Leroi-Gourhan directly on this point, noting that the tool (of whatever kind) is 'a secretion of the Anthropian's body and brain' (TT1, 150/160). It is not just that hands and tool-use put Dasein into contact with other modes of presence (the *vorhanden* and the *zuhanden*) but that, in their wielding of *every* kind of tool, hands bring forth or disclose presence as such. In a statement that might require some further nuance or qualification, Stiegler notes that Heidegger: 'always thinks of the tool as (merely) useful and of instruments (merely) as tools' (TT1, 245/250). Heidegger would indeed relegate the majority of tools, technical prostheses, or technologies to the domain of instrumentality (and therefore pose their essence as being the mode of revealing dominant in modern technology or enframing) but he would also concede the privilege of a more fundamental revealing to those technical prosthetics or modes of tool-use that can be aligned with *tekhnē*, poiesis, art and therefore with the creation and production of poietical bringing-forth. As has been indicated, in Stiegler's account of the technical rootedness of Dasein's relation to time and world-disclosure, no such differentiation can pertain. Stiegler does not wish to maintain any distinction or opposition between an originary poiesis or *tekhnē* and a secondary or derived calculative order of technology as enframing. He puts this in the following terms: 'is not the consideration of *tekhnē* as the originary horizon of all access of the being that we ourselves are to itself, the very possibility of dis-anthropologisingthetemporal,existentialanalytic?'(TT1,262/267). His use of the term *tekhnē* here needs to be understood within the context of his reading of the Epimetheus myth, according to which Prometheus's eventual stealing of the *tekhnē* of fire from the Gods stands as a founding narrative of *all* human technology and tool use. In aligning the human hand with *tekhnē qua* poiesis (as conceived by Heidegger in his two essays on art and technology respectively) and with language and thought conceived as poiesis and bringing-forth, Heidegger, Stiegler implies, remains caught up within a more traditional, metaphysical (one might say logocentric) and anthropological account of the human. Only by accounting for the insertion of a generalised technical prosthetics in and at the originary moment of the giving of Being can, Stiegler argues, this unthought, anthropological and metaphysical legacy in Heidegger be untied and overturned.

It is clear, then, that despite his reliance on Heidegger in *Technics and Time, 1*, Stiegler's reworking of the temporal, existential analytic involves significant deviations and divergences in the way Dasein's relation to technology is understood. In particular he can be aligned more with Jacques Derrida's and Jean-Luc Nancy's thinking of technics, understood respectively as 'originary technicity' and 'ecotechnicity', where what is first and foremost to be thought is, in Derrida's words: 'the ageless intrusion of technics [. . .] of transplantation or prosthetics' within any instance of world disclosure or sensible-intelligible appearance (Derrida 2006: 113/131). What is equally striking, perhaps, is that Stiegler's reading of Heidegger through the lens of Leroi-Gourhan and Leroi-Gourhan through the lens of Heidegger nevertheless leads him to maintain the privileged status conferred upon the hand. From one perspective the hand is a distinctive marker of human specificity in so far as it allows for the advanced manipulation of tools (Leroi-Gourhan). From another, it continues to play a fundamentally ontological role in so far as tool use opens up the ecstatico-temporal horizon of Dasein and discloses a world (Heidegger). Despite his difference from Heidegger, Stiegler maintains a privileging of the hand over the brain. Yet, as has been indicated, his reading of Leroi-Gourhan's account of the upright stance involves an emphasis both on the freeing of the hands for the manipulation of tools *and* on the expansion of the cerebral cortex. His willingness to concede that the 'brain evidently plays a role' is matched by his insistence that: 'it is not the principal determination' (TT1, 145/155). This privileging of the (ontologically disclosive) hand over the brain results in a relative neglect of the question of cerebrality within the overall argument of *Technics and Time, 1*. It is as if Stiegler's reliance on Leroi-Gourhan allows him to treat the evolutionary development of the human brain as an empirical bio-anthropological fact which is an ontic rather than an ontological determination. Stiegler's account of 'epiphylogenesis' and of the way in which human tool use sediments meanings (which in turn make possible our relation to a temporal, historical and symbolic world) is very complex and highly developed. He has very little or nothing to say about the way in which the expanded human cerebral cortex interacts with, or relates to, those sedimented meanings and the ecstatico-temporal horizon that the relation to technics opens up. As was indicated earlier, both Heidegger and Stiegler might have very good reason to place the brain and its cognitive capacities or operations into a secondary or less fundamental position (i.e., their desire to deconstruct a metaphysical legacy of the subject in which the primacy of the brain might play a major role). However, Stiegler's reliance on Leroi-Gourhan means that his philosophical account of technicity and temporality requires the brain and the event of cerebral or cortical expansion, in so far as they play a key role in the advent of the human, whilst at the very same time marginalising them.

The brain and not the hand

The implications of this simultaneous reliance upon and marginalisation of the cerebral in Stiegler can perhaps best be drawn out with reference to the work of another contemporary French philosopher, Catherine Malabou. Since the publication in 2004 of *What Should We Do with Our Brain?* the question of the cerebral has occupied a central position in the development of Malabou's thinking. She also, like Stiegler, develops her philosophy via a close engagement with Derridean thought (she was Derrida's student and collaborator) and with the legacy of Heidegger.[7] Malabou is perhaps best known for her thinking of plasticity, a guiding concept throughout her philosophy which is initially developed in her major works on Hegel and Heidegger and which is subsequently developed in her work on the brain.[8] Malabou's thinking of cerebral plasticity takes recent discoveries from within the realm of neuroscience as her point of departure. Cerebral plasticity here refers to the capacity of neural networks and synaptic connections within the brain to form, reform and transform themselves. Crucially, whilst cerebral plasticity was once thought to be a property of the early developing brain (in the womb and in the first months and years of life), from the beginning of the first decade of the twenty-first century neuroscience discovered that this plasticity in fact continues throughout life, that the brain can go on moulding and remoulding itself at every stage of its development and subsequent ageing. There are a number of ways in which Malabou's philosophical reflection on cerebral plasticity resonates with Stiegler's reflections on technics and his thinking of epiphylogenesis.

First, and perhaps most importantly, brain plasticity for Malabou opens up the horizon of historicity in relation to the human. The fact that there is a permanent possibility of remoulding the brain and its neural networks or synaptic interconnections means that the concept of cerebral 'hard-wiring' gives way to the notion that the brain interacts with, is ceaselessly formed by, and can in turn help to form, the historical world and environment in which we find ourselves.[9] Malabou argues that: 'we can say that there is a constitutive historicity of the brain', noting that, 'what we have called the constitutive historicity of the brain is nothing other than its plasticity' (Malabou 2008: 2, 4/8, 13). So her philosophical understanding of plasticity as that which is able to give and receive form allows Malabou to think a dynamic interaction between worldly projections and engagements and the shaping of neural networks, which allows for some form of freedom or agency in the way in which we engage with the world and form or sculpt those neural networks and their synaptic interconnections. Secondly, this 'historicity of plasticity' is so important because for Malabou, borrowing from the work of neuroscientists such as Antonio Damasio and Joseph LeDoux, the brain and its neural and synaptic interconnections are the site or map of our identity: 'the self is a synthesis of all the plastic processes of synaptic formation at work in the brain' (2008: 58/119). Damasio's idea of the 'neuronal personality' or LeDoux's argument from his book *The Synaptic Self* that 'you are your synapses' (LeDoux 2003: 303) allows Malabou to think the individual and collective being of human beings and their identities, as it were, in terms of brain plasticity and as formable and reformable within the historicity of a finite temporality. Like Stiegler, who has been primarily concerned over the last decade with questions of the individuation and singularisation of the self within the technological and organological 'specificity of our present time' (SM1, 14fr), Malabou is interested in the possibility of harnessing insights into brain plasticity and the 'neuronal self' in order to develop a philosophical–political analysis of contemporary culture. Brain plasticity therefore opens the possibility of thinking: 'a new freedom [. . .] a new meaning of history' (2008: 13/32).

So both Stiegler and Malabou share a concern with the historicity of the human, articulated in the central role accorded, respectively, to the hand/technicity and the brain/cerebrality in the fundamental constitution of what we are. This, in turn, opens the way for each to develop a philosophical–political analysis of our contemporary historical moment. It might be worth recalling here also that, for Stiegler, the primacy of the hand opened up a historical horizon of temporality because of the way in which tool use permitted a sedimentation and retention of meaning traces from an impersonally lived or collective human past. His thinking of epiphylogenesis is a thinking of technical prosthetics as the bearer of meaning traces or of sense which plays a primordial role in the disclosure of an (historical) sensible–intelligible world of phenomenal appearance. Similarly for Malabou, the brain and its synapses are viewed as a bearer of sense, meaning or symbolic activity: 'neurons, by way of their "being-connected" are already available for, and already disposed toward, meaning. In a similar way, meaning or symbolic activity in general, depends strictly on neural connectivity' (2008: 62/127). This in turn makes of cerebral plasticity a 'power to configure a world'. So not only do Stiegler and Malabou think of technicity and cerebrality respectively as constitutive of human historicity, they do so on the basis that tools and brains are viewed as the bearers, or articulators, of a fundamental order of sense and meaning which makes the opening of finite historical temporality possible in the first instance.

Comparing Stiegler and Malabou in this way may now allow for a clearer assessment of the potential strengths and weaknesses of their respective philosophical positions in relation to the primordial status granted to hand and the brain. It was noted earlier that Stiegler's reliance on Leroi-Gourhan meant that he acknowledged the cortical expansion of the human brain as playing a key role in the technological advent of the human. Yet he nevertheless placed the instance of the cerebral in a secondary position and did not develop any full philosophical account of the way in which the expanded capacities of the brain cortex might interact with, or connect to, the epiphylogenetic possibilities opened up by the hand and its manipulations of tools. Malabou's account of brain plasticity might seem to offer a clear way forward in addressing this underdeveloped aspect of Stiegler's argument in *Technics and Time, 1*. Her account of the cerebral in *What Should We Do with Our Brain?* opens the way for a thinking of neural networks and synaptic interconnections understood as the bodily site of a mode of fundamental sense articulation. This site would, along the lines suggested by Stiegler, connect to and 'open up' the world via the hands, tool use and technical prosthetics in general. Helpfully, Malabou stresses that one of the implications of the plasticity and historicity of the brain is that it loses its status as a central command centre, as a site and source of agency. It was noted that Stiegler (and Heidegger) might be suspicious of according the brain a central role precisely because the image of the cerebral as a calculating command centre is implicated in a traditional metaphysics of the subject. Malabou deals with this issue by arguing explicitly that: 'The functional plasticity of the brain deconstructs its function as the central organ' (2008: 35/75).

This is one of the key arguments of *What Should We Do with Our Brain?* and arguably it is a very important one since Malabou needs to show that her reliance on the cerebral *does not* lead her back into a series of more traditional philosophical assumptions. The risk might be that she not only repeats the privileging of cognitive brain activity as the site of a calculative, rational (and metaphysical) subjectivity, but that in so doing she loses track of or forgets the transcendental question of the world and the question of being as posed by phenomenology and fundamental ontology respectively, and that she reverts, despite herself, to a form of unquestioned cerebral empiricism. Her borrowing from the empirical field of neuroscience means that this is a very real risk. For instance, it is arguable that Damasio's idea of the 'neuronal personality' and LeDoux's thinking of the synaptic self (upon which Malabou relies so heavily) are very much framed within an empiricist, brain-centred thinking, which does not account for the fundamental questions of the transcendental and the ontological which are posed within the phenomenological and deconstructive tradition out of which she is working.

Malabou's reliance on Damasio and LeDoux means that, to a certain extent, she is operating with a conception of the brain as site and centre which she then undoes or deconstructs with her thinking of plasticity. Not only does cerebral plasticity deconstruct the brain's 'function as the central organ', it also plays a complex role in what she terms 'historico-cultural fashioning of the self', which occurs in a complex interection between the brain and world, the mental and the neuronal which she characterises as dialectical. Yet, unlike Stiegler, Malabou does not have a developed or sophisticated account of the material, bodily and technical articulations through which the brain might connect to the world, and through which a cerebral dialectic might be articulated. Put simply, just as Stiegler might lack a fully developed account of the brain, so Malabou could arguably be said to lack an account of the hand, tool use, and technical prosthetics.[10]

The hand and the brain, or fundamental organology

What Should We Do with Our Brain? is a short polemical work which is as much focused on a cultural–political polemic as it is on introducing the concept of cerebral plasticity. The question of the implications, difficulties or problems that may arise when one brings the empirical insights of neuroscientists into relation with a philosophical tradition which is preoccupied with a transcendental and ontological questioning of the world and of being is not really developed by Malabou. Nor is the question of the brain's material interconnectedness with the world via other bodily organs posed. At this stage in the development of her thought Malabou is more preoccupied with her philosophical cultural–political polemic than she is with developing the more fundamental philosophical questioning of cerebrality.[11] Similarly it has been argued that Stiegler's work in the wake of the first volume of *Technics and Time*, and during the course of the last decade in particular, has been increasingly preoccupied with a philosophical and cultural–political analysis of, and response to, the contemporary world. As has been suggested here, however, the Stieglerian preoccupation with the hand and with technics comes at the expense of a philosophical reflection on the brain just as Malabou's preoccupation with the brain comes at the expense of a philosophical reflection on how the brain may be interconnected with the world (via hands and other organs). Stiegler is left lacking a full account of the cerebral and Malabou is left with a series of problematic questions relating to the status of the brain and its relation to the world, and to questions of the transcendental and the ontological and their relation to the empirical. Arguably then, as both thinkers come to be preoccupied with political questions, there are aspects of their more fundamental engagement with hands and brains, technicity and cerebrality, which remain in need of development.

In both cases the fundamental, ontological or post-phenomeno-logical thinking of bodily organs remains incomplete or unsatisfactory. This incompletion cannot be resolved within the context of this discussion. It may be enough to suggest in preliminary terms that there is scope for the development of a 'fundamental organology'. Fundamental organology would not, as it has recently been in Stiegler's work, be focused on a politics of individuation or a philosophical analysis of 'the specificity of our time'. Rather it would return to the questions of organs, both brains and hands, both technics and cerebrality, and do so in an ontological register in order to develop further a 'bodily' account of world disclosure. This would be an account which seeks to integrate more fully the insights drawn from an empirical or scientific field (e.g. Leroi-Gourhan, Damasio, LeDoux) with a post-deconstructive, post-phenomenological account of world-disclosure and worldly being or existence. At stake here may be the question of sense or meaning as it is thought by Stiegler in relation to tool use and epiphylogenesis and by Malabou in relation to synaptic or neural interconnections. For a thinker such as Jean-Luc Nancy, a thinker also working within the Derridean and Heideggerian legacy, sense or meaning, and the bodily *relation to* sense or meaning, is the very stuff of our worldly being and shared bodily existence. The Nancian thinking of the 'sense of the world' might well be a resource in taking up the challenge of a 'fundamental organology' in the wake of Stiegler and Malabou's respective meditations on the hand and the brain. To think the two together, and to think the opening of sense in hands and tools, brains and synapses, and to do so without reverting to the prejudices of empiricism, cognitivsm or philosophical naturalism, would be the challenge of such a fundamental organology. This is a challenge that the question of technicity and cerebrality in Stiegler and Malabou places before us but one which has not yet been taken up.

Notes

1. Stiegler supplies a dictionary-like definition of general organology on his Ars Industrialis website: <http://www.arsindustrialis.org/vocabulaire-ars-industrialis/organologie-générale> (last accessed 19 February 2013)

2. Stiegler refers to the concepts he develops as weapons with which a struggle will be waged (SM2, 13–14fr [§1]).

3. Heidedegger's critique of modernity as a mode of revealing being grounded in subjectivity, instrumental rationality and a calculative mathematising projection of the natural world would all be at stake here.

4. See Jacques Derrida, 'Heidegger's Hand', in Derrida 2007: 27–62/ 35–68.

5. For discussion of Derridean responses to Stiegler that are suspicious of the anthropological and metaphysical baggage which might be concealed in his account of the human, see James 2010.

6. Heidegger refers to the need for having a care for the 'dignity' of the event of being and the necessity of 'keeping watch' in relation to that event (Heidegger 1993: 313).

7. See Malabou 2011; Malabou's philosophy takes as its starting point a strikingly original reading of the concept of plasticity in Hegel, see Malabou 2005.

8. Malabou's book on 'destructive plasticity' and brain injury, *The New Wounded* (2012), develops her thinking of cerebrality further.

9. Malabou argues that the organising attribute of brain plasticity its ability 'to configure a world' [*sa puissance configuratrice du monde*] (Malabou 2008: 39/82).

10. *Editors' note:* James is particularly prescient here. While what he describes as the neglect of cerebrality in Stiegler is undoubtedly true of the first three volumes of *Technics and Time*, and indeed of his published work as a whole up until the end of 2012, Stiegler has recently returned to questions of technics and the brain in the drafts of the long-awaited *La Technique et le temps, 4: Symboles et diaboles*, presented in his 2012 doctoral seminar series at the École de philosophie d'Épineuil-le-Fleuriel. These seminars also included an engagement with the work of Catherine Malabou, whom Stiegler criticised for conflating the brain with mind (*l'esprit*), at the expense of an understanding of the relation of the brain to technology, by way of the hand. His renewed interest in neuroscience has played out more fully through the work of Maryanne Wolf, author of *Proust and the Squid: The Story and Science of the Reading Brain* (New York: HarperCollins, 2007). Recordings of these seminars are available to watch online, via Stiegler's website, <www.pharmakon.fr>, though (free) registration will be required.

11. In *The New Wounded* (2012), Malabou develops her thinking of

brain plasticity and the symbolic dimension of the cerebral and neural structures further. She does so more in relation to the question of psychoanalysis and much less in relation to the fundamental world-disclosive or ontological question raised here.

II: Aesthetics – The Industrialisation of the Symbolic

5

Technics, or the Fading Away of Aesthetics: The Sensible and the Question of Kant

Serge Trottein

Which of us, feeling gloomy, on a Sunday afternoon in Autumn –
one of those afternoons when one doesn't want to do anything, but
is bored with doing nothing – which of us has not felt the desire to
watch an old film, no matter what the story, either at some nearby
movie house, if we're in town with a bit of spare change, or at home
if we have a video player, or even, listlessly flipping on the television,
where, although there's no film, just a mediocre soap, or even some old
rubbish, we will still allow ourselves to be carried away by the flow of
images?

    Why don't we turn it off and pick up a book – a book, say, in which
we could find a really good, strong and well written story? Why,
on such a Sunday afternoon, do those moving images win out over
written words in beautiful books?

        Bernard Stiegler, *Technics and Time, 3: Cinematic Time and the
Question of Malaise*[1]

Let us suppose, on the contrary, as an experiment, that on a far
from gloomy summer Sunday afternoon, we don't feel like doing
anything other than reading and analysing Bernard Stiegler's
Technics and Time, volumes 2 and 3, the focus of this essay: as
in Stiegler's story and following his recommendation, we will not
turn on the TV, we will pick up a book instead, and will happen
upon his surprising little story. Is it a really good one, beautiful
enough to drive boredom away (it appears in a section called
'Boredom', '*L'Ennui*')? It may be too early to decide, but meditat-
ing on the sudden appearance of this emblematic story hidden in
the depths of a book about cinematic time may yield insight into
what is at stake in Stiegler's project.

    There is no doubt about it: moving images necessarily win out
over 'written words in beautiful books', which does not mean

than one art is superior to another, that cinema prevails over literature. Even though Stiegler seems to focus on 'beautiful' stories and 'beautiful' books, the question he raises is not a question of *aesthetics* (neither of form nor of content), but of *technics*: the filmmaker, Stiegler writes, only needs to be 'able to exploit the video-cinematographic possibilities' in order to 'attract our attention to the passing images, no matter what they are' (TT3, 10/31). Books may need to be beautiful or interesting, images do not. It is a question of technics, *and* of time – Stiegler's main thesis in this third volume of *Technics and Time* being that consciousness espouses cinematic time all the more easily as its temporality has always been cinematic. Aesthetics or technics? The question seems at the outset to be irrelevant, so invasive is technics or technicity in Stiegler's work.

Or at least in this part of his work, for he also develops elsewhere the theme of the aesthetic war or of aesthetics as a weapon; for example, in his organising of a colloquium on 'The Struggle for the Organisation of the Sensible: How to Rethink Aesthetics?'.[2] But even then, Stiegler's approach to aesthetics is in terms of what he prefers to call 'a general organology' of the sensible (SM2, 83fr [§18]), whose history immediately becomes one with the history of technics: aesthetics is then rethought from the perspective of technics.

Aesthetics or technics?

Nevertheless, in *Technics and Time*, particularly in the volume devoted to cinema, aesthetics is absent – and of course the question will also be whether and why this absence could be more than a simple oversight. In Volume 2, however, aesthetics is not completely absent, appearing incidentally, mainly in section 6 of Chapter 2, as the complement, the enrichment, or even the ornament of a theory which is not his own, but which Stiegler fully adopts, namely Leroi-Gourhan's programmatology. This section, entitled 'Programmes and Aesthetics', begins thus:

> Leroi-Gourhan's programmatology allows not only for the thinking of becoming from the standpoint of transmission as differing repetition, it also *enriches* [emphasis added] this thinking of becoming with an account of the *necessity of forms*, which is confirmed in the behavioural diversity to which the non-zoological transmission

of programmes gives rise. And this dynamic of forms, as the object of *aesthetics*, is also where the technical tendency conjoins with the *who* – and with the who insofar as it desires, as if this conjoining were *erotic*. (TT2, 81/99–100)

Such an aesthetics, whose object is vaguely defined as the 'dynamic of forms', has from the outset great difficulty in finding its place. Stiegler first mentions the 'non-zoological transmission of programmes', but only to add in the very next paragraph that 'even if a sense of the aesthetic as an aspect of group formation and differential reproduction had only been possible since the programmatic event of techno-logic being [. . .] we would still be required to think the aesthetic from the standpoint of the most archaic biological forces'. And he gives two examples, two quotations from Leroi-Gourhan's *Gesture and Speech*, in which one learns that 'there is no fundamental difference separating the crest from the plume, the cock's spur from the sword, the nightingale's song or the pigeon's bow from the country dance' (TT2, 82/100).

And yet aesthetics has to do more specifically with the constitution of human groups, with ethnicity and, above all, with technicity, which ends up absorbing it in its process of exteriorisation, in spite of a certain superiority of the aesthetic over the technical, which Stiegler momentarily seems to recognise, all these reversals taking place in the same section, even in the same paragraph.

> This notion of an aesthetic [that of the nightingale, the country dance, or any human or ethnic group] requires a typological description of programmes as rhythms even more than as memories. [. . .] This development occurs through delegation of functions [. . .] currently [. . .] symbolic, in instrumental programmes – tool, machine, or the industrial complex itself. This phylogenesis of the exteriorisation process precisely demonstrates the principle of the epiphylogenetic evolution of technics [. . .]

Here Leroi-Gourhan is called upon to finish Stiegler's sentence: 'a property unique to the human species [. . .]'. And yet, Stiegler continues:

> The aesthetic's rhythmic programmes are, however, *first those of the body itself*, and more precisely of the body parts responsible for the five senses. It is an Aristotelian strategy that grounds the thought of

the sensible in that of organs whose originary diversity appears to be irreducible. (TT2, 82/100–1)

The thought of the sensible, which seems to define aesthetics in this passage, is, in its 'Aristotelian strategy', still a long way from Kant's aesthetics – which also mentions the nightingale's song and the country dance, albeit in very different contexts – and even farther from Kant's transcendental aesthetics, with its two senses (space and time) instead of the five traditional senses. But at least Stiegler's reference to a 'thought of the sensible' would seem to pave the way for an aesthetics irreducible to the technical dimension of human beings, even if thought in terms of an 'organology'.

Yet more surprises await the reader:

> This aesthetic, which accounts for the evolution and persistence of forms, both of which are *rooted in the technical tendency* [emphasis added], is what enables us to think in terms of 'individual liberty', the 'higher level' of memory in which the symbolic as such, *qua* phenomenon of reflective thought, is worked out. Epokhality is a principle of aesthetic evolution, and it is in this sense that it is doubly articulated, through the technical tendency and through idiomatic singularity. (TT2, 83/101)

Moving from the five senses and the thinking of the sensible to memory, reflection, and the thought of 'individual liberty', from zoology and ethnicity to the technical and the idiomatic, it is indeed difficult to *situate* aesthetics on the sliding scale of this text. Not only do Stiegler's, or Leroi-Gourhan's, pseudo-Aristotelian aesthetics – for it is no less difficult to decide *who* is speaking in this text – affirm the superiority of the technicity in which they are rooted, they are also simultaneously 'a property unique to the human species' *and* barely human:

> This 'physiology of taste', based on the concepts of specific, socioethnic and individual, programmes and memories, which thus must be conceived from the standpoint of zoology, is neither simply 'materialist', in that it doubly articulates the principle of selection, nor simply vitalist, in that in general it breaks such oppositions as animal/human and living/non-living; for the principles of the functional aesthetics are 'drawn from the laws of matter and for this reason [can] be considered as human only in a very relative way'. (TT2, 83/101)

How the principles of functional aesthetics can be drawn from the laws of matter will remain a mystery, at least, probably, until we have analysed Leroi-Gourhan's text, which Stiegler nevertheless adopts as his own. The result is less a deconstruction of metaphysical oppositions (animal/human, technology/zoology, etc.) than a constant shifting from one perspective to the other, in what is not yet cinema, but already a kind of slide show, one might say.

At the end of this section on aesthetics, Stiegler finally distinguishes between three aesthetics: physiological, functional, and figurative aesthetics. But this distinction is immediately erased for the benefit of exteriorisation, which, of course, is just another name for technics. Whereas the aesthetic was supposed to be thought from the standpoint of the most archaic vital tendencies, it now suddenly appears that there is an 'aesthetic becoming', immediately seized and imprisoned however in the network of exteriorisations, where the aesthetic can still be affirmed only in the conditional and ultimately disappears in the synchronicity of one and only one phenomenon:

> The aesthetic becoming conjoins the physiological aesthetic, *the functional aesthetic subjected to the technical tendency*, and the figurative aesthetic (corresponding to the properly symbolic level, i.e. the idiomatic). This conjoining is profoundly anchored in the exteriorisation process to the extent that
>
>> [*here begins the adoption of a citation from Leroi-Gourhan*] technics and language being only two aspects of the same phenomenon, the aesthetic could well be a third [*the last gasp of aesthetics, so to speak*]. We should then have something to guide us, for if tools and speech developed into machines and writing through the same steps and more or less simultaneously, the same phenomenon should have occurred for the aesthetic: from digestive satisfaction to the beautiful tool, to music for dancing and the dance watched from an armchair, there would be the same phenomenon of exteriorisation,
>
> [*and Stiegler completes Leroi-Gourhan's sentence:*] which would extend from the mythogram to the contemporary 'stage of specialisation in which the disproportion between the producers of aesthetic material and the ever-expanding masses of consumers of prefabricated or pre-thought-out art' would increase. (TT2, 83/102, citing Leroi-Gourhan 1993: 275/88–9, vol. 2)

In this world of tools and prostheses, how is a beautiful tool to be found? And moreover, what would constitute a beautiful tool? We may never know, in so far as the disproportion between poiesis and prefabricated art is becoming extreme and the phenomenon of technical exteriorisation has become universal. In the world of technical exteriorisation there is no more room for aesthetics, any more than for art – and aesthetics subsequently, gradually, disappears from the whole second volume of *Technics and Time*. Heidegger, whose problematics Stiegler also largely adopts, was still able, in 'The Question Concerning Technology', to propose or promote art as the privileged domain for reflection on technics, provided it remains open to what he called 'the constellation of truth'. Such a constellation has been extinguished in Stiegler's world, giving way to the unique and overpowering phenomenon of technics as exteriorisation. Even, or especially, in the age of 'prefabricated or pre-thought-out art', the question is no longer: *either* technics *or* aesthetics; *or* now means that technics and aesthetics are fused into the same phenomenon characteristic of our age. In a parallel direction, art has lost or is losing any superiority as an alternative; in other (more Stieglerian) terms, ars (the Greek *tekhnē* or the medieval and classical *ars*) has become *industrialis*.

Thinking the sensible

Given such a configuration, one wonders whether a thinking of the sensible is still possible. It is worth noticing, to begin with, that the context of this question is in many ways similar to that in which Kant found himself in the 1770s and 80s when he was building his critical philosophy. The purpose of Kant's first critical work, the *Critique of Pure Reason*, was to give a definitive answer to the question, 'how is science possible?', in order to avoid metaphysical dogmatism (and 'dogmatic slumbers'), as well as scepticism, itself induced by both endless metaphysical wars and the new empiricism; to put it briefly: in order to tidy up after Leibniz and Hume. However, the question of the conditions of possibility of science presupposes the question of access to the sensible. For the eighteenth century is the age of a rediscovery of the sensible, whether it is called scientific experience, aesthetic taste or moral sensibility. It is the epoch of the invention of aesthetics – Baumgarten coined the term in the second quarter of the century, in Latin, but it was very soon translated into German, and used by the Germans, though in the 1780s they were still the only people in Europe using it, as pointed out by Kant.

Not that there had been no discourse on the sensible or the beautiful before the age of the Enlightenment, but the dominant tendency had always been to erase or obliterate the sensible in favour of the intelligible, and the beautiful in favour of the true or the good. In such a context, it is hardly surprising that Kant's critical work starts with a new thinking of the sensible, which he had to call an *aesthetics*, even though it is itself founded on the erasure or effacement of a previous aesthetics, namely that of Baumgarten. The first paragraph of Kant's first *Critique* is replete with definitions, among them the following: 'The science of all principles of *a priori* sensibility I call *transcendental aesthetic*' (Kant 2003: 66 [A20/B34]). And just after the word 'aesthetic', an asterisk signals the very first note of the *Critique of Pure Reason* (not including the Introduction):

> The Germans are the only people who currently make use of the word 'aesthetic' in order to signify what others call the critique of taste. This usage originated in the abortive attempt made by Baumgarten, that admirable analytical thinker, to bring the critical treatment of the beautiful under rational principles, and so to raise its rules to the rank of a science. But such endeavours are fruitless.

Baumgarten's attempt was in fact already doomed to an inescapable effacement or fading away of aesthetics in so far as his hope was to raise it to the dignity of a science by submitting the beautiful to reason, that is, by reducing the sensible to the intelligible: his success would still have been a failure from the perspective of a thinking of the sensible. In order to establish a place – an irreducible place – for the sensible and for aesthetics, Kant thus decides to give up the idea of aesthetics as a critique of taste, and, by emphasising the distinction between the empirical and the *a priori*, to restore, in his own terms, the ancient division between *aistheta* and *noeta*, the sensible and the intelligible. Hence the term 'transcendental aesthetic', which, he writes, is a 'true science' (Kant 2003: 66 [A20/B34]).

Such a renunciation of an aesthetics of the beautiful in favour of a doctrine of space and time (as *a priori* forms of sensibility), however, will only be temporary. To be precise, it will only last nine years, from 1781 to 1790, from the publication date of the first *Critique* to that of the third *Critique*. Why is the place reserved for aesthetics so unstable? Why is its birth each time doomed to become an abortion? Because thinking the sensible always implies, in one way or another, a certain connection or link to understanding or reason (even in the third *Critique* where it almost immediately appears, in the first paragraph, though with the understatement of a 'perhaps', Kant 2007: 41 [203]). And this link may remain residual or it may prove increasingly invasive, as is the case in the *Critique of Pure Reason*. The problem with thinking is that it is always a matter of ups and downs: in an upward movement, thought raises its objects to a certain dignity by simultaneously submitting them, in a downward movement, to rules, concepts, laws or principles, so that the place initially reserved, the newly opened territory, soon fades away into renunciation and anonymity.

The question of Kant

Have we completely forgotten Stiegler's *Technics and Time*? On the contrary, we are near the heart of the problem. Having seen 1) how Stiegler erases aesthetics from the second volume by making it coincide entirely and irremediably with the technical tendency which is the subject of his work, and 2) how Kant himself had to abandon aesthetics or replace it, at least temporarily, with a new thinking of the sensible, we are perhaps now better prepared to understand the mystery whose exploration, Stiegler writes, is the object of his third volume on *Cinematic Time*. But is the volume on *Cinematic Time* really the third volume of a series of no fewer than five volumes? Does it really belong to the series entitled *Technics and Time*? Its *Avertissement* – warning or 'notice' – is, in multiple ways, a questioning of the book's title, as well as a demonstration of the insufficiency of tertiary retention (a book, for example) to support thinking, even if that thinking is about the cinematic flux of consciousness.

1. First, what is presented as the third element of a series 'could be read as autonomous' (TT3, xi/13), that is, separately from the rest of the series or as its first volume, an *introductory* volume itself full of re-introductions (of problematics from the other two published works), so that it may well be the only indispensable volume of the whole series.

2. Secondly, what characterises in general the whole series, its complex – and certainly non-linear – temporality, is a failure or a lack (*'un défaut'*), not of origin as much as of obviousness (*'d'évidence'*) and especially of connection (a *'défaut d'enchaînement'*), which already raises the question of synthesis – soon to be confronted (TT3, xii/14). The temporality of *Technics and Time*, and maybe its disorientation, is such that instead of coinciding with the global process of exteriorisation which it keeps describing, it breaks down into a series of (re-)introductions to a necessary/needed/obligatory default/lack/absence/deficiency/defect/failure or non-appearance – 'au défaut qu'il *faut*'. So instead of presenting the *last* volume, still unpublished, as the introduction to the series, Stiegler writes:

> Some work remained to be done in order to lead to what constitutes the initial and ultimate motive of the whole enterprise – for the very first version of what should be the last volume of *Technics and Time* was written twenty years ago, and constituted henceforth the initial aim which has never left me since, and one may consider everything preceding it, including the present work, as an introductory discourse to the necessary default, defaults, defections or failures to appear [*au défaut qu'il* faut *à ce qui fait défaut(s)*]. (TT3, xii/14)

The 'initial and ultimate motive of the whole enterprise' may never be known if the publication of the last volume continues to be delayed or if Stiegler decides that it should not be the last volume after all, its title being *Le Défaut qu'il faut*. But it will have been introduced by everything that has been and will have been published in the meantime, and by *Cinematic Time* especially, which is all the more autonomous as it contains its condition of possibility: the remedy for the 'connective fault' (or *défaut d'enchaînement*) noticed after the second volume – a remedy that might well prove to be a *pharmakon*, with the side-effects it implies – is a new re-reading of the *Critique of Pure Reason*, which Stiegler characterises in terms often erased or eluded by the translation:

1. Kant's *Critique* is 'the heart of modern philosophy, the crossroads of philosophy' (*croisée des chemins philosophiques*), but also the 'crux of thought' (*croix de la pensée*).
2. Re-reading Kant's *Critique* offered him, Stiegler writes, 'a kind of familiarity or accord with that region that I had until then

only perceived from far away: the question of Kant' (TT3, xii/14).

With 'the question of Kant' then, we draw very near 'the initial and ultimate motive of the whole enterprise', a proximity that the following sections, in their equally introductory character, are only going to confirm and increase. After the *avertissement*, the prefatory 'warning', comes the Introduction itself, which indeed begins by emphasising the introductory status of the preceding volume: the question of Kant, which needs to be re-opened, is 'the philosophical question of *synthesis*', thought not 'separately from', but just *from* the perspective of an originary prostheticity, and it is this question 'that will constitute the heart of the reflection carried out here through a reading of the *Critique of Pure Reason*' (TT3, 1/18). A little later Stiegler writes: 'we are engaging in a *critical enterprise*, in the still unheard of sense of this word in philosophy, from Kant to Marx and beyond' (TT3, 5/23). It is thus a 'new critique', new because it is a 'critique of present reality', and also because it will be carried out 'in the spirit of the Frankfurt School's "social critique"' (TT3, 5/23).[3] The condition of possibility, however, of such a new critique is yet a further critique, 'a radical critique of the very depths of modern thought *that still remains largely to be undertaken*' (TT3, 5/23 [Stiegler's emphasis]).

The heart of modern philosophy, Stiegler has already affirmed, is Kant's *Critique of Pure Reason*, but he is much more specific in the following passage. The proper object of Stiegler's radical critique is 'the properly critical moment in Kant's thought', that is, the moment in which Kant decides and makes a blind choice, in an occult or hidden manner. This moment is that of the schematism and what leads to it: the transcendental deduction.

At this point, however, there is no longer any question of re-reading the *Critique*, but of *re-interpreting* this double moment as the 'question of the cinema of consciousness, which constitutes an archi-cinema, the concept of which is exposed in the second chapter' (TT3, 6/23–4), after yet another introduction – the first chapter, which deals with cinematic time.

But what makes this double moment 'the *properly critical* moment in Kant's thought'? How can it be at the same time the properly critical moment and the moment when Kant makes a blind choice, a hidden decision? Can it be read or re-read in Kant's text? And if not, how does Stiegler learn of its existence? Certainly not from Horkheimer and Adorno, whose reading of the *Critique of Pure Reason*, he claims two paragraphs later, was 'itself non-problematic and a-critical' (TT3, 6/24), a claim that, at best, undermines his previous endorsement of the Frankfurt School. As Stiegler incidentally acknowledges, arguably letting the cat out of the bag prematurely, the source of all these contentions is none other than Heidegger: 'As Heidegger has contended as far as he was concerned and in his own style [. . .]'.

Such an acknowledgement, however, does not put an end to the multiple introductions to *Technics and Time*, which keep accumulating and piling up: *Cinematic Time*, the first chapter, as we have seen, is one of these, as is its first section about the 'desire for stories' (which again ends with an allusion to Kant's description of schematism as an art hidden in the depths of the human soul), as well as its second section about boredom (used to introduce this essay), and so on. The second chapter, entitled 'Cinematic Consciousness' ('*Le Cinéma de la conscience*'), opens with still more introductions, as though a long period of various initiations or invocations were required in order for the layman to have access to the place where the mystery will finally be celebrated. For example:

> What I call tertiary retention [first invocation!], Heidegger also named [second invocation, here, after many more] *Weltgeschichtlichkeit* (world-historicality) [. . .]
>
>   But we will see here [but not quite here, nor quite now – what suspense!] that this question is the very heart of the *Kantian mysteries* [Stiegler's emphasis] surrounding the question of transcendental imagination. (TT3, 37/67)

This scene is repeated in Chapter 2, section 4, with a new re-introduction of the 'new critique' as a critique of Kant's *Critique of Pure Reason* (TT3, 40/72). But since time is lacking – *le temps fait défaut – il faut*, it is necessary to skip all the other steps of this long initiation into the Kantian mysteries to get at the crux of 'the question of Kant'. And time would lack even more if I were to read Stiegler's re-reading of Kant properly or suitably – *comme il faut* – that is, by re-reading closely and at the same time not only Kant's and Stiegler's texts, but Heidegger's, not to mention Husserl's or Horkheimer's and Adorno's own re-readings of Kant's text. I will therefore limit myself to the following indications.

1. First, the 'facts'. Stiegler, unlike Horkheimer and Adorno, is not going to focus primarily on schematism (a theory rather than 'a concept', according to him), but rather on what precedes it, 'the transcendental deduction of the pure concepts of understanding'. Why? Because this deduction 'exists in two versions', which is true, 'that are significantly contradictory', which is very doubtful, unless one adopts Heidegger's interpretation, which Stiegler fully does, once again in an introductory way: 'We will see here that these two versions, both of which Kant affirms despite their flagrant contradictions [. . .] both stumble precisely over the question of a cinematic consciousness that is constitutive of all conscious activity of which the three syntheses would be the precise operations', 'in tight solidarity', Stiegler adds in the next paragraph, 'with primary, secondary, and tertiary retentions' (TT3, 41/73). Needless to say, this solidarity is not self-evident. It nevertheless allows Stiegler to conclude, before even starting his re-reading, with what will, once again, be the title of the next section: 'Kant's Confusion'. Stiegler does here exactly what he criticised Horkheimer and Adorno for doing at the very beginning of his invocation of Kant: for calling upon a Kantian concept 'as if the concept were self-evident and contained nothing problematic, no critical question' (TT3, 40/72). Perhaps a critique of the new critique would not be so useless after all.

2. Certain details of Stiegler's interpretation would raise still more questions: for example, why does he argue that the triple synthesis of the first edition is later replaced by *two* new syntheses, whereas in fact the second edition keeps on referring to three: the figurative synthesis, the intellectual synthesis, and the transcendental synthesis of imagination (TT3, 46–7/80–1; see also Kant 2003: 164–5 [B151])?

3. The more the analysis bears on the problem of schematism, the more manipulation and prestidigitation, whether in text or image, are foregrounded (TT3, 50–1/86–7). First, Stiegler 'reads' Kant by mixing his text with Kant's exactly as he has been doing in Volume 2 with Leroi-Gourhan about aesthetics, except that this time he is not directly and fully adopting the Kantian heritage, but only doing so through the filter of another adoption, that of Heidegger's thesis (see Heidegger 1997: 50–80 [§§ 16–23]), which he furthermore does not follow in great detail. Not only can such a convoluted way of

reading already be perceived as a kind of manipulation of the reader: the whole analysis is about manipulating symbols and fixing them by way of an image.

Although his critique of schematism may not be quite 'new', Stiegler does at this point introduce a new image: the image of the concept of one thousand. The reason for this new image is to make it more obvious, if need be, that, contrary to what Kant contends, there can be absolutely no schema without an image: 'A number always in some way presupposes a capacity for tertiary retention – whether via children's fingers, a magician's body, an abacus, or an alphanumeric system of writing – which alone can facilitate numerisation and objectification'. And this capacity of quick manipulation, he insists, being a relatively recent development, required a long process of elaboration in the course of human history.

For Stiegler, not only are images, or symbols or tools, indispensable – but, more importantly, someone is needed to manipulate them. Who is this manipulator, this magician? Maybe Kant himself, at least according to Stiegler, since the author of the *Critiques* seems to conceal all that manipulation without which, in spite of being passive, there would be no activity of the understanding – and to add some credibility to Stiegler's thesis, it could be recalled that in the passage where Kant describes schematism as a hidden art, he himself uses the term *Handgriffe* (sleight of hand) instead of *mécanisme*, the word appearing in the French translation of the *Critique* (see the footnote immediately preceding the section on 'Boredom', TT3, 9, 226 n. 2/30 n. 1; Kant 2003: 182 [A140–1/B179–80]).

But isn't the true illusionist or the prestidigitator Stiegler himself? The English version of *Cinematic Time* is here again of little help, even though it seems at least for once to be correct – but it is only correct in so far as it is correcting a very interesting error of the French edition. Look carefully at the new image introduced by Stiegler in order to manipulate the reader so that he or she may more easily adopt his thesis: a succession of one thousand dots filling two thirds of the page (TT3, 50–1/86): you believe him, because, like Stiegler, you are in too much of a hurry, he goes too fast for you to keep up. And count the dots: it can be done quickly, but only because you are actively using a schema, not a synthesis of tools, not tertiary retention – which ends up proving Kant's point, not Stiegler's. The result is: twenty lines of fifty-seven dots for a total of 1,140, whereas the English version of the book displays the proper number of dots, 1,000.

This operation does have something to do with digits and speed, but it is not contemporary technics, rather prestidigitation (etymologically speaking). In fact it is either sleight of hand or blindness – or both at the same time. Whereas the title of Volume 3 of *Technics and Time* seems to promise developments on contemporary art and aesthetics, especially on cinema, the reader is treated to a variety show, after the slide show of Volume 2, by a philosopher performing as an initiator blinded by the mysteries to which he wishes to introduce us. To his credit nevertheless, one might add that Stiegler's emphasis on the question of Kant as the heart of his whole enterprise, inherited from Heidegger's *Kantbuch*, encounters the same limitations and the same blindness (or is it again manipulation?) as are found in his predecessor: why put so much emphasis on the first *Critique* – which begins, as we have seen, with a renunciation of aesthetics – if what one is chiefly interested in is imagination, art, or overcoming metaphysics? In Heidegger (although much more time would be needed to show it, and it would certainly be the occasion of another *show* or another movie with another scenario – Heidegger imagining Kant stepping back from his discovery, and so on), what characterises Heidegger's critique of the *Critique* is its *repression* of the third *Critique* and of aesthetics. By adopting it, Stiegler's new critique repeats this puzzling strategy. To say it in Stieglerian terms then, we still need a critique of the new critique, *une telle critique fait toujours défaut* – unless Kant himself provided it in 1790 and we need only to re-discover or re-invent it.

Conclusion

Meanwhile, a future gloomy autumn Sunday afternoon may necessitate re-reading Stiegler's story in 'Boredom', the second section of *Cinematic Time*. If we cannot find a beautiful book or, because of our laziness, pay only enough attention to it to escape extreme boredom (hence a *malaise*, or rather a *mal-être*, which Stiegler again invokes rather than analyses – an analysis which might be pursued in terms of boredom and entertainment and might this time adopt the *Critical Reflections* of Jean-Baptiste Dubos, the initiator of aesthetics) – if we cannot find a beautiful book or an interesting movie, there will always be, following Stiegler's strategy, the solution of starring in one's own movie: and, if we have to watch one, instead of letting chance decide on its aesthetic quality, let us act, in every sense of the word, in the flux of our consciousness or in writing, in order to overcome 'total laziness', passivity, malaise or mal-être. Such a rediscovery of life and light, by way of some metaphysical miracle where technics finally meets aesthetics, would seem to bring into being something rather than nothing, hope and expectation rather than boredom. But it is doomed to fall back into obscurity and burial: when faced with technics, aesthetics fade away, ultimately unable to forestall boredom, to transubstantiate it for longer than the interval of an evanescent twilight:

> If the film is good, we will come out of it less lazy, even re-invigorated, full of emotion and the desire to do something, or else infused with a new outlook on things: the cinematographic machine, taking charge of our boredom, will have transformed it into new energy, transubstantiated it, made *something* out of *nothing* – the nothing of that terrible, nearly fatal feeling of a Sunday afternoon of nothingness. The cinema will have brought back the expectation of something, something that must come, that will come, and that will come to us from life: from this seemingly non-fictional life that we re-discover when, leaving the darkened room, we bury ourselves in the fading light of day. (TT3, 10/32)

Notes

1. TT3, 9–10/30–1
2. 'La lutte pour l'organisation du sensible: comment repenser l'esthétique?' This colloquium, whose argument explicitly refers to Jacques Rancière's 'sharing of the sensible', was held from Wednesday 26 May to Wednesday 2 June 2004 at the Centre Culturel International de Cerisy-La-Salle, in Normandy. See also SM1, 22. 40; SM2, 16–18fr (§2).
3. The original French text is a little less straightforward: 'dans l'esprit de ce que l'on nomma au XXᵉ siècle la critique sociale, par exemple celle que voulut pratiquer l'école de Francfort'.

6

Experience of the Industrial Temporal Object

Patrick Crogan

If adoption constructs communities, this is also and primarily because technical organs, without which no human society would be possible, are *displaceable*, and because societies can both exchange and adopt them. This is why the conditions of adoption in general are specific to each age and, when they exist, to mnemotechnical specificities. And this is *also* why the question of adoption is indissociable from that of commerce, and therefore of the market.

Bernard Stiegler, *Technics and Time, 3: Cinematic Time and the Question of Malaise*[1]

Introduction

In the third volume of the *Technics and Time* series, *Cinematic Time and the Question of Malaise*, Bernard Stiegler develops an innovative account of cinema as 'industrial temporal object'. The cinema produces for its spectators an experience in time that has had, in Stiegler's view, a profound effect on their experience of time since its inception. This industrial and technical production takes time, the time of the spectator's attention to its unfolding, a time invested in the promise of the experience offered by and expected of the film. If television and video developed commercially by iterating key elements of this cinematic capacity to generate experiences, it is clear that today's ever-expanding digital audiovisual media are also 'cinematic' in this respect. Competition is intense for the time 'investment' of the spectator–user of digital computers, tablets, smartphones, MP3 players, digital broadcast television, videogames, social networking services, not to mention digital cinema (from 3D and other digitally animated effects films, to DVDs and digital video rental services). The notion of the 'attention economy', first arising in the 1990s and being critically reactivated today, strives to come to grips with this aspect of the digital age (Crogan and Kinsley 2012).[2] For Stiegler, understanding how the experience of the cinematic industrial temporal object was crucial in the formation of a 'consciousness market' (TT3, 73) is a necessary preliminary to developing a properly critical response to the 'digital transition' to post-cinematic media.

To this end, this essay will set out to provide some key features of Stiegler's account of the industrial temporal object. It is important to recognise from the outset that this account is not given by Stiegler in a disinterested scholarly manner, but is a key part of the critical project of *Technics and Time, 3* to properly diagnose the contemporary era's growing malaise (as the book's subtitle translates the French *mal être*) or 'ill-being'. The economic mobilisation of the power of cinema and the industrial experience generators that emerge in its wake is central to Stiegler's diagnosis of the contemporary transitional moment.

The economic sphere, which today in its globalising, neoliberal capitalist phase tends to dominate most if not all spheres of individual and social life, drove the mainstream development of the industrial temporal objects of the twentieth century, from cinema and radio and the recorded music business to television and the more recent digital and internet-based forms. The above epigraph from *Technics and Time, 3* points to the intrinsic link to commerce in the artefacts and products of human societies to be found in the development of all communities. It lays out a general condition of societal becoming that is both exploited and, as I will show, brought to a point of crisis in Stiegler's view by the advent of industrially produced and distributed temporal objects. The movement of these displaceable 'technical organs' and, implicitly or explicitly, the techniques for their fashioning of a mode of living, is at the heart of the dynamic through which a collective evolves, renews or reinvents its social, political and cultural identity. The mode of living of the other can be recognised as valuable, legitimate, desirable and even preferable. As Stiegler says, modes can be adopted, and this is why adoption constructs and reformulates communities. The displaceable, tradeable 'technical organs' are key to this power of adoption. Today's globalisation of a consumer-centred identity is both made possible by this power and, in its excessive realisation, threatens the very dynamic through which cultures continue to sustain a viable milieu of individuation for their members.

The distribution of devices and systems providing experiences is the commercial basis of this unprecedented dissemination of a consumerist way of life. The movies were the most influential of these 'technical organs' in the previous century. Consequently, the cinematic techniques of composing experiences that engage the viewer in their unfolding played no small part in the passage toward 'ill-being'. I will examine Stiegler's characterisation of the specificity of cinema as distinct from other technical forms of recording and recounting experience. This is developed through a revision of Edmund Husserl's phenomenological account of temporal experience. This revised, post-phenomenological account of cinematic experience is Stiegler's contribution to the theorisation of film as 'cultural industry', but it is also what sets Stiegler's account apart from the *Kulturkritik* of Adorno and Horkheimer. As I will argue, this influential critique of commercial culture is haunted in Stiegler's analysis by Kant's transcendental schematism. That culture has become an industry should not – indeed must not – lead critical practice away from the potential of the industrial and technical character of cinema. Drawing on Gilbert Simondon's notion of (trans)individuation, Stiegler identifies the challenge and possibility of the industrial temporal object to renew the 'life of the mind'.

The experience of cinema and the cinematic character of experience

For Stiegler, cinema is in a long line of technical forms for expressing and recording experience. These he terms 'mnemotechnics', objects and techniques able to preserve and make accessible experiences which I have not myself lived. These developed as external forms of memory on the basis of the general process of the technical exteriorisation of psychic and physical attributes and potentials that Stiegler, following André Leroi-Gourhan, understands as constitutive of the process of the human's quasi-zoological but also techno-morphological evolution (TT1, 44–50). Mnemotechnical forms such as graphic figuration, body-marking and carving represented a decisive boost to the collective, ethno-cultural dynamics of human becoming inasmuch as cultural identity emerges as a tradition and a legacy on the basis of such mnemotechnical rituals and artefacts.

Experience is, from this perspective, always conditioned mnemotechnically. The mnemotechnical archive has always formed the material substrate for an individual's 'default' orientations to her social and spatial context and to the temporal context of her experience. To be a young or middle-aged woman, of this or that socio-economic class or sub-grouping, or religious community, with a self-identified tribal, ethnic, linguistic or national identity; all of these more or less stable designations have framed an individual's coming to self-consciousness on the basis of previously accomplished, collectively negotiated adoptions of the exterior technical milieu in general and the mnemotechnical archive in particular.

The most profound impact of cinema resides for Stiegler in the way it changed the mnemotechnical 'conditions of adoption' on an industrial and, consequently, global scale. The globalisation of capitalist modernity in the twentieth century is a cinematic as much as it is a cultural and political-economic story. Modernity can be understood as the organisation of the adoption of a 'new connection to time, the abandonment of a privileged tradition, the definition of new life rhythms' (TT3, 92). What this specificity of cinema also reveals, however – and this may be its most important legacy for a renewed critical engagement in the industrial technical transformation of the human 'world' – is that human consciousness has always been 'cinematic' inasmuch as we are always making movies of one kind or another about ourselves.

What is the specificity of cinema? This is for Stiegler always a historical rather than an ontological question. Influenced by Simondon's work on 'technogenesis', which for Stiegler is the most productive basis from which to comprehend the full significance of industrialisation, technical forms such as the cinema emerge in a compositional dynamic between technical, cultural and politico-economic lines of development (TT1, 68–74). With industrialisation, the rapid transformations of the technological sphere tend to dominate the other spheres of existence. A collective stabilisation of the cultural adoption of technical innovations emerges subsequently as the (always provisional) result of a complex dynamic he characterises as a doubling and redoubling. Stiegler develops this notion of the cultural response to technical developments in *Technics and Time, 1* in a critical commentary that insists on historical and material factors that tend to be elided in Heidegger's discussion of a 'history of being' in *Being and Time* (TT1, Part II, Section 3). A concise summary of epochal redoubling is given at the outset of *Technics and Time, 2*:

In *The Fault of Epimetheus,* I demonstrated that the reification of a technical propensity or body of propensities, leading to an altered technical system, suspends the behavioral programming through which a society is united, and which is a form of objective *epokhē* the social body initially tends to resist. An adjustment then takes place in which an epochal intensification (*redoublement*) occurs; this adjustment is the *epokhē's* key accomplishment, in which the *who* appropriates the effectivity of this suspension (i.e., of programmatic indetermination) for itself. Technical development is a violent disruption of extant programs that through re-doubling give birth to a new programmatics; this new programmatics is a process of psychic and collective individuation. (TT2, 6)

The transition from the early years of cinema to the emergence of what became recognisable as the mainstream cinema entertainment form and industry profile around the 1910s–20s has been subject to much inquiry in film studies, with a recently renewed interest in work comparing the beginnings of cinema with the emergence of digital media forms. Stiegler's position corresponds in large part to Mary Ann Doane's in *The Emergence of Cinematic Time,* where Doane states that the passage to mainstream cinema's standardised narrative feature film format, mass produced and industrially organised around capital investment in the distribution of films, was 'overdetermined (culturally, economically, technologically)' (Doane 2002: 159). Stiegler would characterise this passage as the outcome of a cultural and commercial 'redoubling' of the emergent cinematic technics that coordinated it with the overarching 'epochal redoubling' of modernisation. As I have already shown, however, for Stiegler this 'overdetermination' led not only to a meta-stabilisation of what the 'cinema' was to people around the world but to the destabilising of ethno-cultural becoming globally.

For Stiegler, the cinema quickly coalesced into a singularly powerful form for entraining the spectator's consciousness to the unfolding of its duration. This account of the correspondence between the temporal flows of film and consciousness is developed by Stiegler in a critical re-reading of Edmund Husserl's phenomenology of internal time consciousness (Husserl 1991). Stiegler employs while also modifying Husserl's notion of a 'temporal object', namely, an object before consciousness, which is always 'intentional', that is, is always conscious of or toward something.

While every object is temporal to the extent that it exists in time and is subject to the universal law of entropy, a 'temporal object' for Husserl designates a particular kind of phenomenon of consciousness. This is one that takes time to be constituted and cannot be apprehended immediately as a recognisable, unified phenomenon. Husserl's chief example of such an object is a song or a melody, which takes time, the time of conscious attention in the act of listening, to be constituted as complete.

Film is such a temporal object, says Stiegler, one which weds the flow of consciousness to its production of a perceptible experience in time. Its ability to present to the viewer a compelling illusion of reality is based on its automatic, mechanical capture of details of the spatial and temporal dimensions of exterior appearance. This is undoubtedly central to its early success and its rapid expansion into a major industrial media form in the early twentieth century.

Stiegler characterises the specificity of cinematic technics as emerging from the exploitation of the conjunction of these spatial and temporal recording and playback systems. The cinema's extraordinary power emerges as the quickly understood expectation that it is able to generate two 'co-incidences' (TT3, 11). The first of these is the photographic coincidence of past reality, of past and reality. This is the 'real effect' of the capture of a past space-time in front of the photographic apparatus identified by Roland Barthes in *Camera Lucida: Reflections on Photography* and André Bazin before him in 'The Ontology of the Photographic Image' (Barthes 1980, Bazin 1967). In a similar vein, subsequent accounts of the indexical character of the cinematographic sign have stressed this sense of the capture by the camera of what was there before it at the moment of exposure.

The second coincidence Stiegler identifies unites the 'flux' of the film's unrolling in time with the flux of the spectator's consciousness (TT3, 11–12). The Husserlian framing of this account of cinema's effect on the spectator constitutes Stiegler's unique contribution to film theoretical efforts to specify the cinematic medium's experiential character. This uniting of temporal fluxes is the result of the mechanical production of the illusion of movement from the capture of still images, a process that records duration comparable in effect, if not in procedure, to the phonogram's recording of sound – itself later to be wedded to the cinematographic through synchronised sound technology in the re-tooling of mainstream cinema in the late 1920s.

The flux of the spectator's consciousness can merge with that of the film's unrolling inasmuch as both operate to constitute experiences out of the permanent flow of time. Stiegler credits Husserl's proposition of a special kind of memory or 'retention' operating in conscious perception as what best explains the way events and experiences are formed as discrete phenomena. Husserl calls this 'primary retention' to differentiate it from the more common notion of memory as the recollection of past experiences ('secondary retention'). Primary retention works to extend the present moment of intentional consciousness over the duration of the temporal object by retaining in each ensuing moment of its duration a trace of the preceding moment. This allows consciousness to constitute the melody as temporal object via a dual retention and anticipatory 'protention' that at each moment of hearing projects the coherence of the melody based on what is retained of the 'just-past' moments (and their protentions). A comet trail of retained impressions containing retained impressions enables consciousness to hold open a 'broadened now' long enough to constitute, at the sounding of the final note, the complete melody as a single object of consciousness (TT2, 201).

For (the early) Husserl the interplay of primary and secondary retention explains our consciousness of time. Husserl's account of primary retention is important in Stiegler's view for the way it allows one to distinguish between present perception and the reflective and projective workings of the imagination in the production of dreams, phantasms and memories recalled to consciousness. Husserl's account is problematic, however, to the extent that it insists on a definitive opposition between the present perception of phenomena and the workings of consciousness's ongoing review and refiguring of its experiences. For Stiegler, these reworkings of prior experiences are always at work in the present. The constantly developing synthesis of these phenomena that constitutes the persisting dimension of consciousness across or beneath particular perceptions filters, anticipates and otherwise selectively conditions those perceptions. This explains how primary retention is able to retain part but not all of the previous moment in its threading together of moments into a single 'broadened now'. Selection criteria must be operating in this reduction of the perceptible, and these must emanate from the synthesis of consciousness derived from the processing of past experiences. The just-past moments are reduced and retained on the basis of a synthesising instance that both recognises and anticipates – or rather, anticipates the recognition of – the temporal object of perception. For Stiegler it is not that there is no difference between immediate perception and recalled perceptions, but that they can never be absolutely opposed, and that moreover they compose together both the experience of the present and the ongoing development of consciousness as continuity beneath momentary perceptions.

This critique of the absolute separation of primary and secondary retention, and consequently of perception and imagination in Husserl's early phenomenological enterprise, is developed in the last chapter of *Technics and Time, 2: Disorientation* and recapped in Stiegler's analysis of cinema in *Technics and Time, 3*. The goal there is to show how the cinema radicalises the question of the role of exterior mnemotechnical forms in the production of experience. Exterior objects which express and preserve interior retentions and protentions such as paintings or statues are, for Husserl, only other objects available to conscious perception but play no intrinsic role in forming the temporal quality of phenomenality. If Husserl had thought about the new industrial experiential mnemotechnics like the cinema and the phonogram – already well known by the turn of the century when Husserl was developing his account of internal time consciousness – perhaps the separation of exterior 'images' of the past and interior retention might have become a major issue for his phenomenological project. For the reasons noted above, Stiegler promotes all such exterior objects to the status of 'tertiary retention' in order to insist that these mnemotechnical artefacts have always been the material substrate of the collective knowledge about experience that has conditioned the formation of each individual consciousness always already situated in its historical and ethnocultural milieu. The interplay of primary and secondary retention inside consciousness unfolds in this exterior context with which it forms a complex of individual and collective becoming.

The cinema is an exemplary instance of tertiary memory form for Stiegler. It provides compelling sequences of images that unfold in time to constitute coherent experiences able to pass into the memory of spectators. Moreover, Stiegler argues, as the 'discovery' of the famed Kuleshov effect demonstrates, the cinema corresponds in a hitherto unprecedented way with the psychic mechanism of present perception. The Kuleshov experiment, whereby the same image of an actor's face was shown following different preceding shots, is understood in film history to have demonstrated for the Soviet montage school that the viewer 'sees' the shots in a film based on their combination. The exact same shot of the actor Mosjoukine's face emotes hunger, sadness or desire to the spectator according to the preceding shot with which it is paired (a bowl of food, a dead child, a woman reclining on a couch). The way a particular perception is formed from the threading together of retentions and protentions of the just-past moments is demonstrated by the Kuleshov effect. This 'cinematic effect' is not only made at the movies, but is what 'ceaselessly produces a particular consciousness, projecting onto its objects everything that has preceded them within the sequence into which they have been inserted and that only they produce' (TT3, 15). That is to say that for Stiegler the Kuleshov effect not only reveals the specific power of cinema to articulate the living present with fictioned experience, it also exemplifies a 'generalised cinema' of perception in the way it connects 'disparate elements together into a single temporal flux' (TT3, 15). It is both the instantiation of Husserl's consciousness of internal time and its post-phenomenological revision.

The cinema's capacity to wed consciousness to its unfolding belies the absolute opposition between perception and imagination, between primary and secondary retention. For Stiegler, this was a 'phantasm' of phenomenology's desire to ground its eidetic propositions concerning the structures of experience in analysing a present phenomenality free from any subjective distortion (TT3, 16). Like the phonogram which is its major precedent in the industrial production of temporal experience – and one which Stiegler implies it might have profited Husserl to pay closer attention to in his account of how a melody is heard by a listening consciousness (TT3, 20–1) – a film can play and replay the exact same temporal sequence of perceptible phenomena. The fact that each replaying of the same recording constitutes a different experience, a different temporal object, demonstrates that perception is selective and changes as consciousness changes with new experiences, including those accessible via the exterior mnemotechnical archive. A second hearing of the same record changes in no small part because that record has already been heard and played its part in consciousness's ongoing synthesis of experience. What has been 'heard' before colours each new listening of exactly the same record.

As industrially produced temporal object, the cinema magnifies the power of tertiary memory forms to provide compelling experiences able to constitute memories for individuals. It ushers in an epochal shift in the mnemotechnical regime with global consequences. Industrial forms like the cinema and the phonogram produce different 'temporal objects' according to the different circumstances and collective conditions in which individuals come to perceive them. It is also the case, however, that they offer the potential to 'synchronise' the experiences of individuals on a mass, indeed global scale through the dissemination of standardised, identical temporal sequences. This potential is key to understanding the rise of the 'culture industries' and the centrality of the industrial media forms to the dynamics of technocultural globalisation in the century of cinema and beyond.

Adoption, identity, ill-being

Through selective shooting and, almost immediately thereafter, the editing together of diverse recordings into a unified sequence, film doubled the operations of consciousness in the production of perceptions in time. In what Doane calls the composition of an 'articulated time' from out of the recordings of different times and places, film fictioned experience industrially (Doane 2002: 186). The coming together of the main ingredients of what became known as 'Hollywood' in the 1910s standardised the production of this duration synthesised by the cinematic technics of 'co-incidence'.

A veritable cultural reprogramming of the default orientation to spatio-temporality resulted. Cinema amounts to a process of the selective assembling of primary retentions, a process engaging consciousness's secondary retentional sphere of imagination and desire. In doing so it plays its part in the selections and projections which provide the protentional horizon to the flux of consciousness that coincides with its unfolding. Mainstream cinema realised an industrial model of entertainment that engaged the viewer precisely because it was quickly adopted as a structuring of conscious experience (TT3, 63). It provided for 'billions of consciousnesses worldwide' a schema of consciousness, that is, a complex of perception, recollection and recognition unfolding in time.

Stiegler acknowledges the convergence of his account of cinema's power with Adorno and Horkheimer's analysis of the reification of consciousness by the mass mediated, commercial takeover of culture by the capitalist culture industries structuring the leisure time of the masses (TT3, 35–6). For Adorno and Horkheimer, the key to this reification is the capacity of the mass media to divest individuals of the ability to form their own schemas for understanding experience. They refer here to Kant's doctrine of the schemas operating between the *a priori* concepts of the understanding and sensible phenomenality. For Kant these mediate between the sensible and the intelligible, between the empirical realm of lived experience and the transcendental concepts that provide the grounds for making sense of the manifold of the sensible. Adorno and Horkheimer suggest that capitalism has deciphered the 'secret mechanism' of the schemas of the understanding, and unlocked the very coding of experience (TT3, 37).

In developing an alternative to the way this reference to Kant subtends arguably the most influential Marxist, materialist account of mass media culture, Stiegler conducts a lengthy critical reevaluation of Kant's schematism and its place in the dynamics of the mind's synthesising of sense from out of sense perception. An adequate account of this is beyond the scope of my discussion here. Stiegler's disagreement with the *Kulturkritik's* reliance on Kant to explain the true nature of the power of the culture industries is, predictably, with the resort to a transcendental explanation of the power of the industrial, technical production of experience. While the description of the systematic exploitation by capital of the new exterior forms of producing and recording experience is compelling, it is always a question, in Stiegler's view, of technical, and in particular, mnemotechnical conditions. 'Hollywood has become the capital of global schematism', he argues, 'because cinema is the *technical* adoption of unifying representations and phantasms' (TT3, 103). In other words, Hollywood functions as a systematic mobilisation of a systemic shift in the mnemotechnical conditions of individual and collective cultural becoming. It has not unlocked the universally valid secret of human being. There is no such transcendental secret of human understanding or experience. These latter, and the possibility of inflecting, contesting and reforming them, must be thought of in composition with technical and mnemotechnical conditions. Consequently, the question of the human – so long as it remains feasible to approach this bio-technogenesis as a question of the human (and not as the human in question) – must be posed in historical and not transcendental terms.

For Stiegler, the systematic exploitation of the cinematic technics' systemic shift in the mnemotechnical conditions of adoption was most fully achieved first in America because it answered the need for a new mnemotechnics of adoption of the new, modern American national identity. The adoption of American values was largely transacted through Hollywood – America requiring precisely a culture of generalised adoption as a means of generating a collective (labour force) from the multitudinous and diverse immigration flows fuelling its modernisation. Subsequently, or in fact simultaneously, this vision of an adoptable way of life animated the American-led globalisation (replacing older European nationalist and imperial models left behind by the new arrivals). Stiegler cites a comment by American journalist Upton Sinclair from 1917: 'In cinema the world has been unified, that is, Americanised' (TT3, 87).

The paradox of this globalisation of an American identity is noted by Stiegler when he points out, for instance, that people around the world grew up invested in the 'phantasm' of the conquering of the American frontier through victory over the threat of the savage (and undemocratic) Indians (TT3, 107). This political and moral mythologising of forced relocation and subsequent systematic eradication of the various original inhabitants of the continent has been lived out in the imaginaries (and play-time) of children around the world as a shared collective vision of cowboys and Indians. Like Deleuze, who argues that what he designates as the large form action-image that predominated in classical Hollywood cinema invariably told and retold the story of the birth of America (Deleuze 2005: 152), Stiegler identifies a powerful politico-ethical legitimation performed by Hollywood and disseminated globally. 'We are all Americans', he explains in *The Ister* (2004).

Identity will always have its paradoxes, however, because it is always fictional, that is, fabricated from the resources provided by exterior technical forms. Stiegler draws on Gilbert Simondon's master concept of individuation to argue that identity is not a stable ground from which already constituted individuals encounter each other. It is better understood as a project and a projection that works to unite individual and collective in an ongoing dynamic. The collectively negotiated adoptions of the exterior technical milieu described above that condition any individual's entry into and adoption of a particular ethnic or cultural identity are only ever meta-stable, historically and technically contingent shapings of this mutual dynamic. In Stiegler's re-reading of Simondon the 'transindividuation' of psychic and collective identity passes through the technical and mnemotechnical archive which constitutes the permanent substrate of its accumulated 'preindividual reality'. The preindividual is the condition of the projection of an identity that is always imagined rather than assured:

> But this adoption process rests on the possibility – opened by epiphylogenesis (i.e., by technical memory) – of gaining access to a past that was never lived, neither by someone whose past it was nor by any biological ancestor. The process requires access to a false past, but one whose very falsity is the basis of an 'already-there' out of which the phantasmagorical inheritor can desire a common future with those who share this (false) past by adoption, phantasmagorically. (TT3, 90)

This is why Stiegler can say that human consciousness, formed compositionally with technical memory, has always been in a way cinematographic. It has always relied on exterior, technical forms of retention to produce its life as project(ion) of individuation. It selects from amongst fabricated recordings of the past, constituting out of these a present according to a desire for a projected synthesis, a unity of consciousness in and as an I that individuates itself as part of a We. It adopts this unified trajectory of the flow of its experience as its identity, having performatively fashioned it out of the phantasms reanimated from pasts not lived (but accessible via tertiary retention). And it projects this phantasmagorical future unity by fashioning new concretisations, new technical forms to communicate and preserve this desired, animating fiction of achieved identity.

When the cinema assumes a key role in the new industrial reprogramming of identity, it brings about a decisive shift in the scale and nature of the mnemotechnical determination of the conditions of adoption, a shift which leads to an 'epochal' crisis of the dynamics of individuation. 'Adoption, which was once largely determined through politico-religious rituals, can as a result of its mutability be subjected to logistic calculations hegemonically controlled by marketing systems and media forces' (TT3, 92). As discussed above, this hegemonic control was initially directed toward the recruitment and collective organisation of the workforce of modernisation. As productivity increased and the problem of oversupply grew – culminating in the Wall Street Crash and the ensuing Great Depression – the 'logistic calculations' shifted focus increasingly toward the coordination of consumption with the demands of industry.

As preeminent entertainment form of its day, the cinema enters into these calculations in the 1920s–30s via its mobilisation by the expanding 'professions' of advertising, public relations, industrial design, consumer engineering, and so forth. Its power to industrially manufacture experience and to project phantasms of identity is systematically conjugated with the production of needs and wants for the way of life we now call consumer identity or consumer culture.

This systematic exploitation of cinema's (and industrial media's) systemic potential to merge consciousness with a designed sequencing of temporal objects has led to today's developing epochal, global(isation) crisis. An excessive synchronisation of individual minds has disrupted the dynamic of individuation through which the becoming of individuals iteratively modifies the synchronising tendency of collective cultural adoption:

> The individuation of the *I is* that of the *We*, and vice versa, even though *I* and *We* differ. It is because of this that an adoption of the same temporal objects can happen across masses of individual consciousnesses, synchronising their flux. But in *this* case, as we will see, it is not at all clear that there remains a metastability such that *I* and *We* can differ both over a long time and dynamically, that is, such that they can continue to differ *and* to individuate themselves while remaining both different and convergent. We might therefore fear that an entropic process might result from the industrial synchronisation of the time of consciousness. (Stiegler TT3, 94)

The focus on the capture and channelling of attention toward the coordination of consumption with production is key to Stiegler's analysis of this 'industrial synchronisation of the time of consciousness'. This analysis has parallels to Jonathan Beller's account of the attention economy in his reconsideration of the *Kulturkritik* in *The Cinematic Mode of Production* (Beller 2006). If this recruitment of the consumer's attention closes the loop of neoliberal capitalism's 'empire' of global exploitation in Beller's account, for Stiegler it also contains the very seeds of its and our demise. This is because in his view the 'success' of efforts at channelling and coordinating the movements and motivations of individual consciousnesses is leading to the erosion of the very dynamic of individual and collective becoming.

The proliferation and penetration of post-cinematic media forms in more and more spheres of everyday experience exacerbate a situation that Stiegler identifies as a becoming 'septic' of the transindividual dynamic linking individual and collective (TT3, 74).[3] The saturation of commercially-oriented mediation generates a form of 'synthetic cognition' as Stiegler characterises it in *Technics and Time, 2* (TT2, 188). This is fashioned out of a fixing or 'cramming' (one sense of the French *se caler*) onto each moment of the 'living present' of consciousness the production of adoptable desires in an ever-spiralling arms race between real-time, pervasive, attentional technics (TT2, 188/219). The fabrication of perceptions for the purpose of synchronising memories and thereby the protentional mechanisms for anticipating future experience and future identity becomes increasingly apparent as a commercial engineering of desire and experience. If the coordination of experience is always the basis for the metastable cultural programming of transindividuation, the consumerist-oriented 'hyper-synchronising' of consciousnesses represents the destabilising of the balance between individual and collective individuation. A 'short-circuiting of individuation (of temporalisation)' results from this effort to overdetermine the global course of the flux of consciousnesses (TT3, 99).

As elaborated across several works in Stiegler's activist critique of contemporary technoculture, a discrediting of and disbelief in cultural identity, social reality and even individual existence emerge as a consequence (SM1, SM2, TC, DD1, DD2). This is what characterises ill-being as a pathological detouring of psychic and collective individuation. In Stiegler's account the symptoms of this are everywhere apparent, from the discrediting of politicians and the political process, to the absence of a collective resolve to act in the face of global financial and ecological crises, to the emerging extremisms and fundamentalisms that rush into the void left by the collapse of enlightenment logics of western social 'progress' or of the Weberian 'spirit of capitalism', and the acting out by disaffected individuals and groups in their name (EHP, 33–44; AO).

Conclusion: make movies

I argued above that Stiegler acknowledges the significance of Adorno and Horkheimer's account of the culture industry's reification of consciousness via the industrial design and global dissemination of schemas of experience. Stiegler challenges, however, their ascription of this reification to capitalism's discovery and exploitation of the secret transcendental mechanism of the mind's schematisation of experience. He maintains that the schemas of experience are not universal but must be analysed through consideration of the historical development of technical and mnemotechnical forms that alter the milieu in and through which humans associate, communicate and record their experiences. This resonates with his fundamental philosophical claim that the question of human being is a question that concerns a becoming composed with technogenesis. As a mnemotechnical form which transformed the 'conditions of adoption' of cultural and individual becoming, it is crucial to apprehend cinema's effects on the accumulation of collective archives for preserving and projecting cultural identity. This is in his view a necessary preliminary to developing a better critical approach to the industrial mass mediation of experience that arose in its wake.

Stiegler's work has progressively articulated the terms of a 'new critique' conjugated with an active adoption of the systemic potentialities of the technical milieu. Characterised variously as 'organological' and 'pharmacological', this renewed critique of technoculture is elaborated in several places. It advocates a consideration of the human as always constituted in historical, contingent combinations with exterior, technical 'organs' enabling its individual and collective survival and development. This fundamentally prosthetic, dependent condition, one of 'being in default' of an essence (TT1), means that the technical milieu is a necessary, inevitable component of human becoming, both poison and cure – *pharmakon* – for its projection of identity (TC, Stiegler 2012a).

This is why in recent texts about the significance and the legacy of cinema today Stiegler argues for the need to 'make movies' and to 'get behind the camera' (Stiegler 2011a, 2012b). This is consistent with Stiegler's interest in a strategic revival of the 'amateur', originally an enlightenment figure whose knowledge of and competence in a given practice proceeds from a love and a commitment not reducible to commercial, economic or instrumental ends. In *Taking Care of Youth and the Generations*, Stiegler states that this new figure of the amateur is at the heart of the politics of the Institut de Recherche et d'Innovation, the research centre he founded at the Pompidou Centre (TC, 200 n. 31). Moreover, the promotion of a critical and practical engagement with digital technoculture is central to the activism of Ars Industrialis.

As part of a critical program, developing a 'working knowledge' of the production of experience has the potential to revitalise the circuits of individuation, a potential which Stiegler notes has re-emerged with the digital remodelling of established broadcast media production and reception regimes (TD, 221). It is not only that getting 'behind the camera' represents an effective way to unmask the synchronising techniques of mainstream cinema (and the industrial experiential media that have succeeded it) – a rationale which best characterises the project of the 'political modernist' experimental filmmakers of the 1960s–70s (Rodowick 1994). Drawing inspiration from Abbas Kiarostami's enigmatic *Close Up* (1990) and its significance for the film-loving Iranian society in and for which it was made, in 'Faire du cinéma' Stiegler appeals to the critical value and necessity of working with as well as on – and not against – the technics for fabricating such compelling experiences. This will enable 'us' to better understand the nature and necessity of 'our' cinematic consciousness. It is in this way that a better adoption of the (post-)cinema's systemic, industrial mediation of experience can be opened up in and as a way to remake a credible future.

Notes

1. TT3, 91
2. This is not least in the wake of Stiegler's polemic in *Taking Care of Youth and the Generations* concerning the damage to established pedagogical and cultural forms of cultivating attention by its unchecked commercialisation. See also Stiegler's essay (2012a) in the 'Paying Attention' issue of *Culture Machine*.
3. Elsewhere Stiegler will characterise this as a pollution of the psychosocial milieu comparable to the other environmental pollutions generated by industrialisation (TC, DD2).

7

The Artist and the Amateur, from Misery to Invention

Martin Crowley

For Bernard Stiegler, contemporary existence is characterised by what he calls 'symbolic misery'. In the volumes bearing this title, Stiegler depicts a situation in which, as he puts it, aesthetic experience has been replaced by conditioning, producing alienation and anomie on a massive scale. The political stakes of this situation are considerable: if it is to be remedied, Stiegler argues, this will require a reinvention of political life, inasmuch as this depends on the shared occupation of an aesthetic dimension (broadly understood as the public circulation and validation of affect and its expression, and roughly equivalent to Rancière's 'distribution of the sensible').[1] This chapter considers Stiegler's diagnosis of this state of affairs, and the responses he formulates as part of a struggle against it. After an initial sketch of the scale of the problem as Stiegler sees it, I will look at the mechanics of its production, its consequences for artists and audiences, and the pharmacological fightback Stiegler proposes.

Stiegler describes our contemporary aesthetic existence as a 'catastrophe'. By 'aesthetic experience', he understands participation in the production and circulation of symbolic material as the expression and development of sensory and intellectual existence. In his account, this experience has been subjected to the 'seizure of control of the symbolic by industrial technology, in which the aesthetic has become both the weapon and the theatre of economic war' (SM1, 13fr). The central feature of this 'war' is the substitution of conditioning for aesthetic experience, in a process which suborns the desires of audiences, feeding them standardised fare organised according to the imperatives of marketing. While this description is openly indebted to Adorno and Horkheimer's account of the 'culture industry', it goes beyond this by seeking to identify the mechanisms which make this substitution more than an exploitative accident: such an understanding is, for Stiegler, the condition of an effective fightback.[2] This understanding also produces a particularly bleak vision, however. As we will see, the mechanisms by which marketing techniques have colonised the consciousness of mass audiences are, in Stiegler's account, integral to human consciousness as such. This explains the extent of the problem: accessing the structure of consciousness, marketing campaigns have been able to produce an aesthetic conditioning which 'substitutes itself for aesthetic experience in order to make it *impossible*' (SM1, 21fr).

We will shortly look in detail at the mechanisms that make this possible. For now, what needs to be stressed is that mass exclusion from aesthetic experience is indeed, for Stiegler, catastrophic. It leads on the one hand to the death of art (not just its end in a Hegelian sense, but its death: the end of aesthetic experience; TT3, 169/169; SM2, 152fr [§33]), and on the other to mass immiseration. For Stiegler, exclusion from aesthetic experience as he defines it equates to social exclusion. He describes the division in question as follows:

> an immense part of the population is today deprived of any aesthetic *experience*, entirely subjected as it is to the aesthetic *conditioning* in which marketing consists, which has become hegemonic for the vast majority of the world's population – while the other part of the population, which still has such experiences [qui expérimente encore], has resigned itself to losing those who have sunk into this conditioning. (SM1, 20–1fr)

This exclusion forms what Stiegler calls a 'ghetto', functionally if not geographically at the heart of contemporary civilisation; those outside the ghetto need to take its agonies seriously, as 'the devastation produced by the *aesthetic war* which the hegemonic rule of the market has become' (SM1, 22fr). This is the catastrophe: massive exclusion from aesthetic experience, on the one hand; the complacent illusion of such experience, on the other.

The challenge that Stiegler poses is that of a new politics of aesthetic experience: not the aestheticisation of politics or the politicisation of aesthetics, but the realisation that aesthetic experience, at the heart of shared existence, is a fundamental political question, to the extent that politics is concerned with the quality of this existence. 'That the world of art has abandoned political thought is a catastrophe', he writes; and reciprocally, 'that the political sphere has abandoned the question of aesthetics to the culture industries and the sphere of the market in general is itself catastrophic' (SM1, 18fr). Accordingly: 'The fact that aesthetic and symbolic life henceforth finds itself hegemonically subjected to the interests of industrial consumption must be placed at the very heart of artistic and political practice and thought' (SM2, 186fr [§39]). This renewed engagement with aesthetic existence is, for Stiegler, the condition of a struggle against its overwhelming commercialisation. Pharmacologically, he identifies the weapons of this struggle in the very technologies deployed to foster the immiseration he denounces. Far from lamenting 'humanity's industrial and technological destiny', he says, it is a matter of 'reinventing this destiny' (SM1, 23fr). Those technologies devoted to aesthetic control and conditioning also make it possible to imagine 'a new epoch of the circuit of desire': the task is to begin the struggle for *new modes of existence* (SM2, 104fr [§23]). The rest of this chapter looks at how Stiegler moves from his bleak diagnosis to a vision of this struggle in the aesthetic sphere; to this end, it is now necessary to look more closely at the mechanisms responsible for the 'symbolic misery' he describes.

Hypersynchronisation

The industrial exploitation which Stiegler sees as characterising the contemporary relation to circuits of symbolic exchange is, he argues, neither accidental nor simply an alienation. If this exploitation has proved so effective, this is because human consciousness is already so structured as to make its industrialisation eminently possible. By definition, for Stiegler, human consciousness constitutes itself by means of those technical prostheses which enable it over time to form ongoing syntheses of sense and memory data. The industrial production of such prostheses accordingly enacts a fundamental modification of the human relation to the world. There is nothing nostalgic about this position: rather, his proposition is that each new technical form articulates the human relation to the world differently; and these articulations can be more or less sustaining, more or less damaging. The relation between human consciousness and prosthetic supports is itself qualitatively neutral; it is the industrialisation of these supports that gives Stiegler cause for concern. If he considers this industrialisation so damaging to the forms of consciousness it produces, this is because it blocks the process through which individual consciousnesses are constituted by moments of individuation in long, transgenerational circuits of symbolic exchange articulated across successive technical forms of external, inherited memory, a process Stiegler calls 'epiphylogenesis'.[3] ('Individuation', for Stiegler, is always at once singular and collective: individual and milieu are co-individuated, and this process is ongoing.) For this reason, 'the current prostheticisation of consciousness, the systematic industrialisation of the entirety of retentional devices [*dispositifs*], is an obstacle to the very individuation process of which consciousness consists' (TT3, 4/21).

If the industrial exploitation of consciousness produces such a blockage, this is due to its reconfiguration of what Stiegler, following Husserl, calls the 'temporal object'. As defined by Husserl in *The Phenomenology of Internal Time-Consciousness*, the temporal object is distinguished by the fact that it exists solely as movement through time; the classic example is a melody. The temporal object is aligned so closely with the perceiving consciousness that the latter 'adopts' the former: while I am listening to the melody, the time of my consciousness is the time of the melody (see TT3, 33/33, 61–2). This relation of adoption makes possible the risks which characterise the industrial production of temporal objects: 'Adoption', writes Stiegler, 'can as a result of its mutability be subjected to logistic calculations hegemonically controlled by marketing systems and media forces' (TT3, 92/145). The structural openness of consciousness, articulated via technical prostheses and coinciding with its temporal objects, exposes it to manipulation by commercial interests seeking to direct the operation of these prostheses in order to channel consumer desire towards standardised forms of consumption. This is the process, at the heart of our 'symbolic misery', that Stiegler describes as 'hypersynchronisation'.

The relation between an individual consciousness and its collective milieu is organised for Stiegler by what he calls the composition of two dimensions: the synchronic and the diachronic. A consciousness, he says, 'in the sense acquired by this word during the seventeenth and eighteenth centuries' (TT3, 3/20), is essentially diachronic: pursuing its development to its own singular rhythm. This development also, however, requires occasional synchronisation with the pre-individual, collective milieu, in moments of individuation through which consciousness and milieu are continually re-defined. Thus: 'The *we* is always both, in one and the same movement, but each time singularly [. . .] an apparatus [*dispositif*] for *synchronisation* AND for *diachronisation*, that is to say for *differentiation* in the *composition* of these tendencies' (SM1, 120fr). When new media make it possible for individual consciousnesses simultaneously to adopt en masse the same temporal object, this composition is threatened with decomposition, as the synchronic becomes massively dominant. This constitutes 'hypersynchronisation', in which for example 'The same program can be watched by millions of viewers at the same moment; millions of consciousnesses can be immersed simultaneously in the flux of the same temporal object, subjected to the same effects of belief and adoption' (TT3, 121/184). Synchronisation itself is not the problem: this is necessary, as we have seen, and is supplied by exceptional moments (festivity, worship, celebration, etc.) in which individual consciousnesses negotiate their distance from the collective milieu (see TT3, 100/155). With hypersynchronisation, this exceptional temporality is dissolved, threatening the very viability of individual diachronic consciousness:

> When people watch the same televised event, at the same time, live, in their tens of millions, indeed in their hundreds of millions, consciousnesses the world over are internalising, adopting and living through the same temporal objects at the same time [. . .] these 'consciousnesses' end up becoming the consciousness of the same person – which is to say, *nobody* [*de la même personne – c'est-à-dire* personne]. (SM1 50–1fr)

Do shifts from broadcasting to 'narrowcasting' and from analogue to digital media alter this situation, by reaffirming the diachronic singularity of individual consumers? In Stieglerian terms, such developments may in fact be seen as exacerbating the problems of hypersynchronisation. Targeted or on-demand programming, especially consumed on mobile devices, offers an illusion of distinction; but, offering such distinction 'in the privative mode' (as we might say in Heideggerian vein), namely as mere isolated consumption of identically-structured material, they equate for Stiegler to the staggered adoption of identical objects and the repetition of identical behaviour. He writes: 'the diversification of channels is also a way of targeting particular consumers [*une particularisation des cibles*], which is why they all tend to do the same thing' (SM1, 26fr). And digital media, which might offer new creative possibilities, are 'perverted' (SM2, 89fr [§19]) by being deployed massively to stifle differentiation and favour standardisation. Literal hypersynchronisation – the simultaneous mass adoption of identical temporal objects via broadcast media – can now be joined by the illusion of its opposite – the freedom of targeted downloads recommended on the basis of individual consumer profiling – to produce the dispersed mass adoption of formally identical objects, the temporality of whose reception is everywhere and each time the same.[4]

The elimination of singularity at work in hypersynchronisation favours the interests of so-called 'cultural' capitalism. Repeating in the realm of consumption the industrial standardisation of production, hypersynchronisation reduces the unpredictability of individual behaviour, which can thus more accurately be translated into the quantifiable data required by financial speculation. According to Stiegler, capitalism, 'having become cultural and at the same time hyperindustrial [. . .] tends to eliminate those singularities that resist the calculability of all values on the market of economic exchange' (DD1, 37/64). This represents a tendency 'to liquidate politics properly speaking' (DD1, 36/63): first by dissolving the power of the state, but secondly and more significantly by destroying 'the psychic and collective individuation process in which singularities are formed and exchanged – this process being itself the experience of one's own singularity, which is to say also of one's *incalculability*' (DD1, 36–7/64). This is Stiegler's 'symbolic misery': standardised consumption of identical objects, alienation from participation in circuits of symbolic exchange, loss of that diachronic dimension necessary to the dialectic of ongoing individuation, and consequent fading of a sense of my existence as incalculably singular. If the industrial revolution alienated the producer, says Stiegler, the hyperindustrial turn does the same for the consumer:

> The loss of *technical* individuation hits the producer: deprived of his skills [*ses savoir-faire*], he loses his technicity. The loss of *aesthetic* individuation hits the consumer: deprived of the possibility of participating in aesthetic activity [*le fait esthétique*], he loses his sensibility. He sinks into anaesthesia, indifference, and apathy. (SM2, 50fr [§12])[5]

Human existence for Stiegler is those processes of individual and collective development which, sustained by transgenerational forms of inherited external memory, are articulated through long circuits of symbolic exchange. Locked into short consumerist circuits of repeated ingestion and expulsion, we are estranged from these sustaining longer rhythms. And if individuation is threatened by the decomposition of the dialectic between the synchronic and the diachronic, tipping the subject of aesthetic experience into the role of anonymous consumer, something similar is happening on the other side of the process of symbolic exchange. For the artist is hardly immune to these developments.

Hyperdiachronisation

For Stiegler, art entails a dialectic between diachronic singularity and collective memory. He defines art as 'the *experience* and the *supporting structure* of this sensible singularity as an invitation to symbolic activity, to produce and encounter traces within collective time' (SM1, 26fr), and as 'what seeks to make possible another experience of time [*faire temporaliser autrement*], so that the time of the consciousness of an "I" [. . .] can always be diachronic, and can set free, by affirming it, the narcissistic, unexpected dimension [*l'inattendu narcissique*] of its singularity' (SM1, 182fr). In contrast to the calculability sought by hypersynchronisation, Stieglerian aesthetic experience produces something authentically unexpected, provoking individuals and groups into measuring the play of their desire against the fabric of external inherited memory, and so sustaining their processes of psychosocial individuation. When artworks work, he argues, artists and audience encounter something that could not have been calculated in advance; and participation in such encounters is vital to ongoing individual and collective development (see SM1, 179–80fr). Answering the question, 'What is an artist?', Stiegler accordingly highlights the particular importance of diachronic singularity. The artist is, he writes, 'an *exemplary* figure of individuation', performing the singular style in which each 'I' continually re-articulates its relation to the groups through whose stock of external memory it exists, and which are themselves composed of these fluctuating re-articulations (SM2, 251fr [§52]). Key to this exemplarity is its relation to time, and especially to unpredictability. The modern artist emerges as a figure when art is distinguished from both artisanal and industrial production, says Stiegler. The historical specificity of this figure resides in its exhibition of incalculability as an exceptional relation to time (a relation of diachronic singularity), just as industrialisation is developing the standardisation of time on a massive scale: 'the artist then becomes a relation to time which [. . .] *does not calculate*, just as everything else is being subjected to calculation' (SM2, 256fr [§54]).

Breaking the dialectic of synchronic and diachronic, the industrialisation of aesthetic experience eliminates the singularity of its subjects through hypersynchronisation; correspondingly, those who produce works of art see their diachronic exceptionality neutralised and parodied as 'hyperdiachronisation'. 'The loss of participation', writes Stiegler, 'is followed by [*a pour corollaire*] the hyperdiachronisation of artistic individuation' (SM2, 264fr [§56]): incalculable exceptionality persists, but only to the extent that it is itself drawn into the realm of calculable commodification. In the early twentieth century, this leaves the artist with two roles. On the one hand, 'the hyper-diachronised expression of the singularity which the industrial apparatus [*dispositif*] synchronising forms of behaviour and sensibility cannot eliminate' (understood preeminently by Duchamp); on the other hand, 'an individual laboratory for aesthetic research and development [. . .] in the service of industrial aesthetics' (the moment of Bauhaus) (SM2, 257fr [§54]). The two names which for Stiegler mark the subsequent development of this situation are Warhol and Beuys. Warhol performs hyperdiachronisation to perfection, 'understands the characteristics of the individuation as loss of individuation induced by reproducibility', but 'does not interrogate its miserable consequences, does not seek in it the possibility of *the other modality* of repetition' (SM2, 149fr [§32]). Against this Beuys, insisting on art as 'social sculpture', seeks to revive a functioning relation of co-individuation between artist and audience, singular and collective. 'Warhol records the consumerism whose misery is told by Beuys', as Stiegler puts it (SM2, 149fr [§32]). Beuys is especially important to Stiegler, in fact, distinguished by his understanding of the stakes of twentieth-century art. For Beuys as for Stiegler, human existence is aesthetic existence: Beuys, in Stiegler's words, 'declares that *every human existence is intrinsically artistic*, and that as such, every man is an artist' (SM2, 108fr [§23]). Stiegler's Beuys also sees that this aesthetic existence is mutilated; and so that what is needed is 'the broadening of art as a process of psychosocial individuation fighting against the loss of participation' (SM2, 175fr [§38]). Beuys's slogan, 'Everyone an artist!' thus declares the project of social sculpture as 'a fight for organisation of the sensible' (SM2, 181fr [§38]). By the start of the twenty-first century, however, even Beuys's efforts have failed. Hyperdiachronisation triumphs, and the artist disappears into the commercial role of designer (culmination of the Bauhaus route) or a brand name. (Beuys's critics, of course, accuse him of both nostalgically denying the reality of this situation and cynically exploiting it.[6]) Faced with this, then, how does Stiegler propose we combat our 'symbolic misery'? The answer lies in fact not with the artist, but with the subject of aesthetic experience; to approach this, we need to revisit the question of time, specifically Stiegler's understanding of repetition.

From artist to amateur

Eliminating the incalculable in favour of predictable investment and return, hypersynchronisation–hyperdiachronisation favours repetition in its toxic form: the compulsive repetition of the same. Different objects, but always identically structured, as so many novelties. Just like Hegelian 'bad infinity', this is the 'bad repetition' of mere accumulation. What is lost in this, for Stiegler, is repetition in a strong sense: 'the *experience* of repetition, that is to say of repetition *as learning*', or 'repetition *as practice*' (SM2, 145–6fr [§31]), the 'good repetition' of growth and change, enabled by repeated encounters with the same object over time; repetition as difference, as Stiegler here points out with reference to Deleuze. (Or the '*other modality* of repetition' Warhol failed to seek.) The proliferating differences of commercial product offer nothing but the same, short-term rhythm; repeated encounters with the same aesthetic object can, on the contrary, let us experience something different each time. And the long epiphylogenetic circuit of these encounters, reconnecting with and differing from the collective milieu, can sustain our existence. This conception of informed repetition constitutes Stiegler's pharmacological weapon against symbolic misery; to understand its possibilities, we can consider the practice of good repetition he identifies in regular encounters with particular works of art through acts of religious worship.

What matters here for Stiegler is the repeated but exceptional encounter with works of art (specifically, within his culturally contingent scenario, music and painting). In these encounters, he argues, the repetition of a structure (here, of an act of worship) supports a practice of ongoing individuation. Returning to the same image, say, worshippers mark their own differentiation from the collective milieu, and the continual development of this milieu. The object of this experience is encountered as always to be re-encountered; in Stiegler's words, the painting says: 'You'll have to come back and see me again, otherwise you won't see me' (SM2, 140fr [§30]). In this return lies the possibility of a practice, an informed exposure to processes of individuation over time. What Stiegler here calls the painting's 'future anterior' offers open-ended subjective and collective transformation; for this, such moments need to be both regular and exceptional, and the objects they present characterised not by the fake novelty of the commodity, but by the difference disclosed in repetition as learning.

Now, if such a practice is to challenge and even overcome symbolic misery, who might be identified as its subject? To answer this, Stiegler revives the very figure whose aesthetic experience was thought to have been killed off by industrial repetition: the amateur.[7] For the one thing at which the amateur excels is a creative practice of informed repetition. Stiegler's use of the term 'amateur' draws heavily on its etymology: a non-professional, of course, but crucially, one who practises an art, or a skill, out of love for the activity. (An 'amateur d'art' is an art lover, for example.) And this love is sustained by a practice of repetition. Practice and love are linked, for Stiegler: drawing on arguments advanced during the eighteenth-century 'Querelle de l'Amateur' (which saw Diderot and other non-aristocratic 'hommes de lettres' opposed to a nobility insisting on the superior taste conferred by its station), he suggests that love of a work of art is tied to intimate, informed contact with its composition.[8] Sketching paintings in a gallery, or learning pieces of music, constitute practices of repetition which enact the dialectic of singular and collective in evolving processes of individuation, sustained through regular engagement with particular technical forms. In arguing for the value of this kind of amateurism, Stiegler also draws on the meanings of 'répétition' in French: in addition to its English senses, the French term refers to repetition as the regular practice required to learn an activity (it is the term for a theatrical rehearsal, for example). Amateurs repeat – practise scales, say, or copy out a score – in order to maintain their love of these objects; through such repetition, they continually become who they are, in relation to and distinct from the collective inheritance they are engaging with, perpetuating, and modifying (see SM2, 153fr [§33]). The alienation of the consumer, de-skilled by the replacement of aesthetic experience with commercially driven conditioning, is here opposed by that knowledgeable practice whose loving repetitions allow the development of processes of individuation characteristic of such experience. In opposition to the hypersynchronisation of the consumer and the hyperdiachronisation of the artist-as-brand, this practice offers the diachronic experience of singularisation through repetition, in which I can *'come back to myself, but as the difference in my repetition'* (SM2, 266fr [§56]). Just like the artist's constitutive diachronisation, this experience is a matter of time:

> Then I have the feeling of taking my time again [*de reprendre mon temps*], being taken up again *by* time, *in* time, feeling myself individuating myself anew, once again, *da capo*: far beyond the pleasure principle, the joy of aesthetic experience lies in experiencing time, and experiencing time in multiple ways. (SM2, 266fr [§56])

Artist and amateur, united in their informed practice, step out of the time of standardisation, and deploy repetition to open the possibility of the unexpected. Here we have in its simplest form the circuit of participation in symbolic exchange from which Stiegler argues the great majority of contemporary citizens have been excluded. No work without an audience, and no audience without a work: 'For all of this to take shape [*Pour que tout cela se constitue*]', writes Stiegler, 'there has to be a circuit and a form of participation' (SM2, 39fr [§10]).

The key phrase here, though, is perhaps the first: 'For all of this to take shape'. For we have seen that, in Stiegler's account, artists and audience are equally excluded from such a dynamic – which is to say, this dynamic is no longer operative. The classic figure of the amateur was killed off by technologies of reproduction, especially the recordings which made it possible to enjoy music at home without being able to play or read it. (Stiegler's witness to this change is Bartók, who laments the passing of an informed amateur audience: see SM2, 34–5fr [§8].) And the artist has been dissolved into either designer or brand. It cannot therefore simply be a matter of reviving a neglected practice: this fantasy would lack any historical understanding. What is required is much more extensive:

such practices require the reconstitution of an organological community opening the possibility of a new 'distribution of the sensible',
which has to be organised, and as a form of social organisation. This
means that it can only be the result of a battle which is still to be
fought – and which requires 'an art which is still to be invented' *as well
as a reinvention of the question of politics*. (SM2, 154fr [§33])

Stiegler is here quoting Beuys: the amateur who will invent this
new art actualises the creative potential affirmed by Beuys's
'Everyone an artist!', in the political reinvention Beuys called
social sculpture. 'Everyone an amateur!', perhaps.

If there is hope, as Orwell's Winston Smith might say, it lies in
the amateurs. But how are these to form themselves, as part of this
reinvention of politics? At the heart of this is a historical understanding of technics, what Stiegler calls the 'general organology'
made possible by his understanding of technical forms as historically specific epiphylogenetic sedimentations. Invocations of 'the
work of art', or 'the artist', can seem airily ahistorical; but there is
little of that here. Art for Stiegler steps out of time, appeals to a different temporality: but this characterisation is historically specific,
defined against the standardised time of industrial production.
Moreover, any future challenge to the hyperindustrial standardisation of aesthetic experience is made possible by the ongoing
technical developments which are the history of human existence.
Stieglerian amateurs are technical and historical all the way down.
That alternative experience of time, which Stiegler describes as the
joy of aesthetic experience, is accessed 'prosthetically, by means
of instruments, works and other forms of organised trace' (SM2,
266fr [§56]). As we have seen, the commercial exploitation of
temporal objects is not an aberration, in Stiegler's account: it is a
structural possibility opened by the relation between finite human
consciousness and the prostheses through which this consciousness becomes what it is. But this structural possibility also offers
a chance. Pharmacologically, it is the technicity of the artwork –
which, in its most recent guise, killed off the amateur and replaced
experience with conditioning – that offers the possibility of a
meaningful encounter.

Stiegler's pursuit of what he calls 'an ecological politics of spirit/
mind [*de l'esprit*]' (TT3, 78/123) commits him to discerning the
possibilities for creative individuation offered by any given technology, over and against the forms of exploitation it may also make

possible. This is his distinctive pharmacological turn. Nothing is in itself poisonous; anything can be more or less harmful or beneficial, depending on how it is deployed. Accordingly, there is no reason why the structures that support hypersynchronisation cannot become the basis for an informed practice of participatory repetition, as 'the possible devices of a new epoch of repetition as production of difference' (SM2, 43fr [§10]). This other, good repetition is to be sought within the same forms currently deployed to foster the bad repetition of consumerist addiction. Although they understood that machinic repetition was the question that needed to be addressed, then, in Stiegler's view Duchamp, Warhol and Beuys did not yet grasp what his approach reveals: namely, as we have seen, that an *other modality* of repetition' can and must be developed within and against this machinic repetition (see SM2, 149fr [§32]). Stiegler's analysis thus seeks to avoid both the easy assimilability of Warhol (the flickering effectiveness of minimal ironic displacement) and the Romantic–Modernist nostalgia of Beuys (still longing for organic social fusion), to find ways to reinvent cultural and technical forms of alienation as tools for transformative practices of repetition.

The techno-pharmakon

This, then, is the pharmacological approach Stiegler's amateurs will need to adopt. He finds an ideal example in the history of the phonograph. In the first place, the phonograph significantly changes the reception of temporal objects: with its introduction, 'the identical repetition of the same temporal object became possible' (SM1, 79fr). Part of the industrial homogenisation of consciousness, the phonograph tends to promote repetition without difference, that unexceptional repetition of the same which eliminates the unexpected in favour of calculable output. It also, as we have seen, deskills its users, and brings to an end the era of the amateur, duly replaced by the consumer. But considering the phonograph not as disaster but as *pharmakon* allows Stiegler to attend to those ways in which the new technology was also used – by Charlie Parker, and also by the same Bartók who regrets the passing of an amateur audience – to invent new ways of listening, as part of the processes of creative individuation they engaged in with their new audiences (see SM2, 33–5fr [§§7–8] and 48–9fr [§11]). The phonograph thus offers a specific case of pharmacological redeployment. In what appeared to be no more than the mechanical repetition of the same recorded sequences, musicians such as Parker or Bartók, a performer such as Glenn Gould, and listeners alert to these possibilities, could discover a new aesthetic experience. The key to this is the realisation that nourishing artistic repetition – repetition as productive of difference – is also to be found here: that 'in each new occurrence of the same temporal object of recorded music, the phenomena appearing to musical consciousness are different' (SM2, 48-9fr [§11]). Finding the cure within the poison, and giving the example of the circuit of production and reception invented by modern jazz, Stiegler argues that this experience opens 'a brand new epoch of repetition', sustaining processes of singular and collective individuation (SM2, 49fr [§11]).

In Stiegler's pharmacological analysis, then, the very technology that appeared to have killed off the amateur turns out to offer the possibility of this same figure's re-emergence. Just as the phonograph is becoming obsolete, it serves for Stiegler as an instance of that process of creative reappropriation which has been opened again, in his analysis, by digital media:

> the analytic capacities of digital machines are again renewing musical language and practice, and making it possible to imagine the transition from the age of consumers [. . .] to the age of amateurs, who *love* because, in their own way, by forms of practice which cannot be reduced to habit, they too *open out* and, in so doing, are opened. (SM2, 32fr [§7])

Stiegler is, of course, far from suggesting that digital media *per se* lend themselves to active amateur production rather than alienated consumer reception. But he does claim that they open this possibility anew. 'A new connection between the poles of listening and creation, but also between watching and listening, between the body of the receiver and the body of the sender, such a connection would be one of the most interesting results of the machinic turn of sensibility', he writes. Indeed: 'it is the condition for the reinvention of the figure of the amateur' (SM2, 39fr [§10]). It is by using these technologies as informed amateurs that we might come to contest the standardised consumerism they are also serving to impose. There is a discipline involved: such contestation is only available once I have acquired sufficient understanding of the technology in question to extract it (if need be) from its existing commercial context and redeploy it, inventing myself in the acts of individuation it makes possible via the circuit of symbolic participation. The close connection between production and reception Stiegler identifies in digital media also shows how becoming an amateur might be thought of as becoming an artist: informed participation in such media invariably means creating and circulating material. The reinvention of the amateur might, on this account, coincide with the reinvention of the artist. At which point, it becomes helpful to expand the definition of the artwork (along the lines of Beuysian social sculpture, indeed) such that it names pre-eminently these acts of circulation, and the forms of individual and collective development they sustain. This experience will be familiar to many; it is happening every day, from bedroom to club to street to bedroom, in technical rearticulations of collective memory. For Stiegler, it is an indispensable part of the struggle against exclusion from participation in those processes of symbolic production which, as he might put it, make life worth living. It is hardly sufficient, of course: the existence of such amateur creativity within the bosom of symbolic misery, and especially its spectacular role as an escape route into conspicuous consumption, suggest its fragility when unsupported by a coherent project of aesthetic politics – itself integrated into a fundamentally different economics (as delineated in Stiegler's *For a New Critique of Political Economy*), and major aspects of social policy, most notably education. Stiegler is scarcely suggesting that digital amateurs can make the change by themselves. But he does sense that their activities might not be part of a significant pharmacological contribution:

> Reality TV, karaoke, sampling, a million French people composing music on their computer according to a study recently published in France, house music and DJs, whose capacity for invention is well known, blogs: these are symptoms of the symbolic misery induced by the loss of participation in which a possible future can therefore also be heard knocking at the door. In this misery, there might perhaps be something revolutionary. (SM2, 57fr [§14])

In 1986, Jean-François Lyotard gave a paper at a colloquium organised by Stiegler on 'Nouvelles technologies et mutation des savoirs'. The conclusion to this paper shows how well Lyotard had grasped the stakes of the work Stiegler was beginning to undertake. Referring to 'the new mode of inscription and memoration that characterises the new technologies', Lyotard expresses his concern regarding the new forms of domination these may be thought to harbour, asking: 'Do they not impose syntheses, and syntheses conceived still more intimately in the soul than any earlier technology has done?' On the other hand, he continues (in a somewhat pharmacological vein), might they not be thought to offer, 'by that very fact', new forms of resistance? He concludes: 'I'll stop on this vague hope, which is too dialectical to take seriously. All this remains to be thought out, tried out' (Lyotard 1991: 57). Refusing Lyotard's melancholy, Stiegler would doubtless want to formulate these new possibilities not as resistance, but as invention, as he puts it in *États de choc* (EC, 132–4 [§30]). That aside, however, this is indeed Stiegler's project: to think through this problem, and to try out responses. Between Lyotard's vague hope, impossible to take seriously, and his own intimation of some kind of revolutionary possibility.

Notes

1. Stiegler both acknowledges and differentiates himself from Rancière's analysis: see SM1, 19.
2. For Stiegler's acknowledgement and critique of Adorno and Horkheimer, see TT3, 35–40/65–72.
3. Stiegler's discussions of epiphylogenesis can be found throughout his work, from the first volume of *Technics and Time* onwards. In *De la misère symbolique*, see for example SM1, 77–80. For a succinct explanation, see James 2012: 67–8.
4. I thank Suzanne Guerlac for this question of mobile and on-demand programming.
5. Stiegler is here developing Simondon's analysis of the industrial alienation of the worker. See also SM1, 101–3, and Simondon 1989, esp. 78–82.
6. See, respectively, Buchloh 2000 and De Duve 1988.
7. On Stiegler's revival of this figure, see Crogan 2010.
8. On this, see Stiegler's forthcoming *Mystagogies*, Ch. 3.

III: Psychoanalysis – The (De)sublimation of Desire

8

'Le Défaut d'origine': The Prosthetic Constitution of Love and Desire

Christina Howells

> My wound existed before me. I was born to embody it.
> (Joë Bousquet, 1941)[1]

> Man does not exist.
> (Jean-Paul Sartre, 1960)

> Man is a recent invention [. . .] in the process of dying.
> (Michel Foucault, 1966)

> Woman [*La femme*] does not exist.
> (Jacques Lacan, 1971)

> And what if we were already no longer humans [*des hommes*]?
> (Bernard Stiegler, 1994)[2]

These provocative claims have all been notoriously misunderstood, both wilfully and through sheer intellectual philistinism. The 'humanism' (or post-humanism) of Sartre, Foucault and Lacan is undeniable and deeply unconventional. The political implications of their (anti-) humanism are multifarious and cannot be conflated or synthesised. Indeed, post-metaphysical critiques of the subject appear to have coincided with the collapse of the political and of the attempt to think the political. Derridean deconstruction too has been similarly criticised for its susceptibility to recuperation by nihilism on the one hand and messianism on the other (Žižek 2008: 225): if democracy and justice are always 'to come' ('*à venir*') then the risks of either a resigned acceptance of the status quo or alternatively an idealisation of some future utopia may indeed prove fatally tempting to some less subtle political thinkers and theorists.[3]

It is against this well-established intellectual backdrop of the second half of the twentieth century that Bernard Stiegler is operating when he comes – relatively late – to philosophy in the 1980s and 90s. Stiegler's own discussion of the human is far from naïve: he is not advocating a 'return to the human' or a 'rediscovery of the subject' as if the hard-won anti-metaphysical progress of the previous decades could, or should, be simply cast aside. On the contrary, Stiegler's work takes the deconstruction of the humanist subject as a given, and uses it as his starting point for a sophisticated re-examination of both humanism and subjectivity. Unlike Sartre and Derrida, Stiegler is not a conventionally trained philosopher, but the breadth of his philosophical reading is remarkable and he is, like them, deeply concerned with the political implications of all philosophical positions, including those apparently unconnected with society or politics. Stiegler's most striking ideas (technics and pharmacology) develop on the one hand out of the Heideggerian notion of *tekhnē* and on the other from the Derridean *pharmakon*, both of which, of course, have many centuries of philosophical history behind them, traceable in the latter case to Plato, and in the former beyond Plato and Aristotle to Xenophon. This essay will examine the relationship between *tekhnē* and *pharmakon* in Stiegler's thought and consider their implications for love, desire and ultimately politics.

Le défaut d'origine

As we saw in the Introduction to this collection, one of Stiegler's major claims is that *tekhnē* has been, from the outset, subordinated philosophically to something other than itself, be it to *épistémē* or to so-called 'living memory', whereas it is, in his account, rather the very condition of philosophy and indeed of experience itself. We looked briefly at the various Greek myths of origins as they are explored by Stiegler in *Technics and Time*: these suggest that human specificity is not a matter of some essential or intrinsic quality as it is for animals but rather arises from the gift of fire, stolen from the gods by Prometheus to compensate for the forgetfulness of his brother Epimetheus, and later the gift of Pandora, the first woman, who arouses human desire. Man, as we saw, is thus constituted by something that lies outside him, resulting not in a human nature, but paradoxically a human *lack of nature*. And this is precisely what Stiegler means by *technics*: humans are 'prosthetic', they are constituted by their interactions with something external to them, they do not have a pre-existing essence, indeed they have no qualities to call their own.

So far things may not sound too good for Man: he was initially forgotten, he is not self-contained, he does not have a human nature to call his own, he is incomplete and condemned to ever-unsatisfied desire. But all this will be turned around by Stiegler, in an operation of *qui perd gagne*, loser wins, that we will doubtless recognise from Sartre among others. It is precisely the absence of essential qualities that makes us human, unbinds us, opens us up to the world outside and ultimately to other people. Our missing origin is the 'défaut qu'il faut', the de-fault that is necessary for us to avoid being caught up in the trammels of self-identity and stasis. But, just as for Sartre this lack of essence is both what permits human freedom *and* what produces the terrifying anguish freedom brings in its wake, so, for Stiegler this 'défaut' (fault or default, a term which he prefers to Lacanian 'manque' or lack, TC, 217n.5/199n.1) is both what allows us to invent ourselves *and* what condemns us perpetually to seek something outside and beyond our (non)selves to complete and indeed constitute us.

Pandora too is, and brings, a mixed blessing; she is what Stiegler will later call a *pharmakon*. She creates a desire that she cannot satisfy, but she also introduces a whole range of other torments for hapless mankind in the jar (or box) which she opens in her boundless curiosity, inadvertently allowing the torments to escape and roam the world. This will be familiar to all those who know their Greek mythology, or indeed who have read *Technics and Time*. What I propose to do in this essay is to trace the sometimes unexpected consequences for love and desire that Stiegler draws from the myths. I will refer in particular to the three volumes of *Technics and Time* as well as to *Aimer, s'aimer, nous aimer* (2003)[4] and to *What Makes Life Worth Living: On Pharmacology* (2010).

The 'General Introduction' to *Technics and Time* describes how the conjunction of epimetheia and prometheia produces the double effect of *elpis*, anticipation, both hope and fear. According to the myth, *elpis* was the one quality that didn't escape from Pandora's jar when she removed the lid. In later years, and especially in the New Testament, *elpis* has the primarily positive meaning of hope, but in Classical Greek, *elpis* is an ambiguous quality, involving projection of the future, be it good or evil. This is the sense Stiegler prefers to retain. And what makes us human is precisely this ability – indeed this condemnation – to anticipate, to look forward to the future, and in this future, of course, to foresee our death. Stiegler writes of *epimetheia* and *prometheia*:

> It is their inextricability that gives mortals *elpis*, both hope and fear, which compensates for their consciousness of irremediable mortality. But this counterbalance is only possible given the de-fault of origin in which Epimetheus's fault consists – namely, the originary *technicity* from which *epimetheia*, idiocy as well as wisdom, ensues. (TT1, 16/30)

When we look to the future we are confronted by our inevitable mortality: the ambiguity of *elpis* ensures we greet this with hope as well as fear for the moment of its coming is always unknown.

Stiegler is classically Derridean (or indeed Sartrean) in his insistence on the human *défaut d'origine*. What is specifically and originally human is a *défaut*. This *défaut* ensures that we are not, and can never be, self-identical. There is nothing (other than a de-fault) to be identical to. There is no originary self. It is only through Prometheus's gift that we can constitute ourselves as human: the gift of fire is at one and the same time the gift of *technics*, that is to say an inescapable interrelation with exteriority, and the gift – for it is a gift – of *desire*. For technics and desire are themselves inseparable. Fire as technics is what enables and compels our prosthetic constitution: it is only through interaction with what is outside us that we can exist as individuals; and this constitutive intertwining with the world and other people condemns us to perpetual desire: we need others simply to exist as human, there is thus no possibility of autonomous existence, free from the fire of desire. We have no interiority to retreat to, even were we to wish it. 'The being of humankind is (to be) outside itself' (TT1, 193/201). Desire is both inescapable and perpetual – nothing and no one can ever satisfy it. If we are lucky, this desire will be sublimated in love; if we are unlucky, it will be frittered away in ever less satisfying pleasure.

In *Technics and Time, 2* Stiegler stresses the positive aspect of our absence of qualities. It is precisely what saves us from fixity: 'The most terrifying thing would be for *The* Human to exist. It does not exist, any more than language. Humans exist, and all language is already languages' (TT2, 162/187). If The Human existed we would be condemned to stasis and sclerosis, and therefore, we might argue, humans, in their perpetual creativity and inventiveness, would not exist. So the existence of The Human would be paradoxically incompatible with the existence of humans.

Stiegler's discussion of the prosthetic nature of the human is developed in several different domains. These include aesthetics, and in particular the question of the image; psychoanalysis, for an understanding of the way in which other people are constitutive of individual interiority; and science and technology in so far as these explore the human relationship to the outside world. All of these are vital for a full appreciation of Stiegler's philosophical analyses of love, desire and politics.

Stiegler draws on Lacan's and Barthes's rather different theorisations of the image to develop his argument about the originally prosthetic nature of the human, though his familiarity with Lacan here, as elsewhere, seems relatively limited. 'According to Lacan', Stiegler writes, 'the mirror constitutes the human as such' (TT2, 25/37–8). It is true that Lacan contrasts what the human infant can make of its reflection in the mirror with what the chimpanzee can do, but this is a far cry from the mirror 'constituting' the human. On the contrary, and as Stiegler recognises, the mirror only ever returns a deceptive image of self-identity and unity to the viewer. Our reflection in the mirror helps us construct ourselves as egos, but not as subjects. Stiegler is right in a sense to refer to the 'de-fault of self' that appears in the mirror: what is reflected is not the human subject but rather the fictional self, solidified in an illusion of autonomy and inertia. Of course, Stiegler is free to make whatever use he wants of Lacanian theory, but perhaps he would have done well to make his (mis)reading more explicit. What clearly attracts Stiegler to this aspect of Lacan is the way in which he understands the mirror precisely as a constitutive prosthesis: 'The mirror institutes an interminable *maieutics of the self* in which exteriority is constitutive' (TT2, 26/38). But Stiegler fails to distinguish sufficiently between the Lacanian self and the subject (in part, perhaps, because of his own wariness of the term 'subject' and preference for the term 'individual') and ends up attributing to Lacan a more positive view of the mirror stage than his thinking really allows and not recognising that the mirror stage is only ever a means of human alienation. The mirror stage is indeed essential to the establishment of the ego, but it is thereby a means of subjective *misrecognition* which involves a spurious identification with an *imago* (Lacan 2006a: 76/94–5). (We shall see an analogous misunderstanding of Lacanian *jouissance* later in this essay.)

Barthes, however, fares much better. Stiegler is again drawn to Barthes's discussions of the image, in this case the photograph, as a constitutive prosthesis of the human. But unlike Lacan, Barthes does not describe the image as a form of alienation: on the contrary, the photograph precisely reveals us to ourselves in the truth of our mortality. Unlike the mirror, the photograph captures a moment that has already passed, and ultimately brings us face to face with death. Whether the photo is of someone still living or already dead, whether infant or adult or old man, the photo necessarily re-presents a moment of life that is no longer present, and it forces us to think about the relationship between temporality and human mortality. The photo provides undeniable evidence that time passes, never to return; it allows me to see what could not otherwise be experienced, 'the past life of the other, his death as other, and, by projection, my alterity in my own photographic mortality: "every photograph is this catastrophe"' (TT1, 266/270). Stiegler explains this further in *Technics and Time, 2*. '*The past is present* in the photograph' (TT2, 14/25).

> This always-already retarded specularity allows me to see myself, here, in my photogram, death. The subject of the photograph, captured by the lens, is mortified, deadened: objectified, 'thinged'. It becomes phantomic. In the exemplary experience of the subject's (re-)seeing him- or herself photo-graphed, in the pose's wake, late, too late, death comes into view. (TT2, 17/28)

Stiegler takes over as his own Barthes's comment 'All photos are this catastrophe' (TT2, 17/29; Barthes 1980: 148), and again 'all photography is a narcissistic and thanatological catastrophe' (TT2, 114/137), arguing that 'the photograph contains an objective melancholy binding time and technique together' (TT2, 18/30). He describes Barthes's *punctum* (the special affect which makes some photographs so powerful) as a 'wound in the spectator [...] a work of mourning' (TT2, 19/31). Stiegler develops this analysis considerably further in his discussions of cinema, and notably of Fellini's film *Intervista* where two actors watch themselves in a film they made together over thirty years earlier. Youth is present only in the past, and the future is clearly marked out as death.

The image then is analysed by Stiegler as intrinsic to the constitution of the human through a process of internalisation and externalisation: contrary as it is to conventional ways of thinking,

I do not pre-exist my self-recognition in the image (be it of myself or of another): there is rather a kind of aesthetic dialectic between me and my reflection through which I am brought into existence (albeit in a mode of alienation for Lacan). Further exploration of this idea is one of the major contributions of *Aimer, s'aimer, nous aimer* (2003), where Stiegler discusses what he calls 'primordial narcissism',[5] which he distinguishes from the more familiar psychoanalytic notions of primary and secondary narcissism.[6]

Primordial narcissism and sublimation

In Freud, primary and secondary narcissism refer respectively to infantile self-love or libido, and to the later redirection of libido from its external objects back to the self in adult life. But Stiegler uses the term 'primordial narcissism' to refer rather to the indispensable self-love that enables us to love others, and indeed which he argues is itself precisely enabled by the narcissism of a 'we' (a 'nous') and not merely of an 'I' ('je') (L, 38–9/14). He links this form of narcissism to Lacan's mirror stage, as he understands it: it is only in so far as we have a sense of ourselves, and of our self-worth, that we can have a real sense of the worth of others. It is the lack or failure of this primordial narcissism which Stiegler sees as key to many of the current ills of society, for we cannot love others, individually or collectively, if we do not love ourselves and do not have a deep-rooted sense of our own existence (L, 42/18). The alternative to social and personal integration is – in what is perhaps superficially an apparent paradox – a loss of self in the impersonal collective of 'das Man' (L, 80–1/86–7). Primordial narcissism is not, then, a matter of egoism: on the contrary, it is essential to a positive relation between individual and collective and is precisely what fends off social atomisation (L, 48/30). Stiegler uses Simondon's notion of transindividuation – discussed in more detail in *Technics and Time* – which maintains that interiority is a product of our relations with others, to reinforce his own argument about the necessary and indeed desirable dependence of the individual on others for his own personal individuation, which is a dynamic process and never complete. So once again we see that it is our inadequacy, in the sense of our instability, lack of autonomy, and 'défaut d'être', 'my inadequation with myself' (L, 76/80–1), that protects us from the stasis and death of self-identity. We can 'subsist' without individuation and integration, but we cannot 'exist'. The alternative to primordial narcissism is the fatal loss of love and desire (L, 64–6, 82/58–60, 89) and the consequent liquidation of the law and social order (DD1, 12/30).

Desire and love, as we already know from Sartre, Lacan and Jankélévitch, to name but three, are never capable of satisfaction, and this is our salvation as well as our painful condemnation. Desire and love look forward, beyond the present, beyond what we already possess, and yearn for an impossible point of union that would inevitably entail their death and dissolution. Desire is infinite (L, 47/27). I love infinitely. This is of course a fiction: we know from Freud that libidinal energy is limited, but belief in the infinite nature of love and desire is once again a necessary fiction or fantasm (L, 48/28). And such fictions are an essential consequence of our *défaut d'origine*, which is in constant need of supplementing. But no supplement can ever complete us, though we always hope that it will: to give up this hope would be to lose the very 'sentiment d'exister', the feeling of being alive, that is so vital to our humanity. In consequence, these fictions and supplements that keep us full of anticipation are always ambivalent, or what Stiegler (following and expanding on Derrida) will call pharmacological. This is the explicit theme of his 2010 work *What Makes Life Worth Living: On Pharmacology*. Our existence is permeated by ambivalent objects of desire, objects that give our life meaning but which can, by the same token, destroy us. Mythologically, as we saw, the first *pharmakon* was fire itself, quickly followed by Pandora; psychoanalytically, according to the object-relations theory of Donald Winnicott, which Stiegler uses extensively, the first *pharmakon* is the transitional object (or security blanket), which stands in for the mother in her absence and allows the infant to acquire maturity and autonomy gradually, without losing its vital 'sense of being' or alienating itself to some kind of false self which feigns a maturity it does not yet possess. The transitional object, then, is imbued with a power that is no less significant for being ultimately fictitious, and infantile dependence on it is key to the later independence of the adult. In Stiegler's terms, 'the transitional object has a remarkable virtue: it does not exist' (WML, 12fr [Introduction]); that is to say, the object itself exists, of course, but its metaphysical significance goes far beyond its physical being, it represents the 'amour fou' (passionate love) between the mother and child. There are connections here with what Sartre says about mother-love in *L'Idiot de la famille*, where he argues that such unconditional love gives us a feeling of personal value that will be vital for us in later years: mother love gives us a 'mandat de vivre', an illusion of necessity, which, he maintains, is a necessary illusion ('une aliénation heureuse') in the face of existential anguish and contingency (Sartre 1971: 143).

Transitional objects can never, however, save us from the dissatisfaction inherent in our *défaut d'être*, just as good relations with the mother may preserve us from the desperate feelings of inadequacy and pointlessness of a Richard Durn – discussed in Stiegler's *Acting Out* – but cannot ensure that we will ever be truly happy: 'The tendency towards neglect is irreducible: there is not, there has never been, there never will be heaven on earth' (WML, 202fr [§69]). Nonetheless, finitude, unhappiness and loss can be made immeasurably more painful and more noxious by the generalised failure to understand, channel or contain them. Similarly, at a more intimate level, our avoidance of the mask of a false self by no means implies that there is a true or authentic self we might aspire to achieve:

> Does the false self presuppose a true self that would be 'authentic' and 'proper'? Evidently not: it is a transitional self, a relationship which is constructed beyond inside and outside, and which must be thought of as starting from a *pharmacology of the soul*. (WML, 115fr [§37])

There never was any pre-existing interiority (WML, 41fr [§9]), and self-construction is therefore a matter of making something out of nothing, and creating interiority from what we encounter in the world and in other people. It would probably be reductive to describe this as 'making the best of a bad job', but it is certainly possible to make what may seem from the outset to be a raw deal (forgotten by Epimetheus) infinitely worse. And the avoidance of this, we may remember, is what may at first sight seem the rather minimal aspiration of psychoanalysis: the aim of treatment as Freud describes it is to transform 'hysterical misery' into 'common unhappiness' (Freud 1955a: 305), just as for Lacan there can be no 'cure' other than learning to accept the impotence of desire and the failure of love. But even such modest ambitions are vastly preferable to the debilitating pain that can be experienced by the truly unloved and unfortunate in society.

What, then, is the best that can be expected for human life given our originary *défaut d'être*? If it is this *défaut* that stimulates desire, and thereby makes us human, what exactly is the nature of the possible benefits that come from our failure to achieve the stasis of stable self-identity? The simple answer to this is everything 'that makes life worth living' (WML, 123fr [§40]): love, friendship, knowledge, art, literature, fidelity and philosophy (WML, 45fr [§10]). It is desire and our inability ever to satisfy it fully that enables and indeed compels us to continue to be creative: we strive to create ourselves through our (prosthetic) relationship with our environment and other people, and the inevitable failure of this attempt to achieve fulfilment means that it is constantly renewed. Our acceptance of this process is manifest in what psychoanalysis calls sublimation, that is to say the aspect of our libidinal economy that channels desire away from immediate satisfaction towards the higher rewards of culture, sociability and love. But Stiegler (following Winnicott) is not satisfied with the Freudian theorisation of sublimation: Freud rightly recognises sublimation as vital to human autonomy and creativity, but fails to acknowledge the element of dependency that the process involves in so far as love, for example, can never be sated by full possession of the beloved and is thereby also a form of heteronomy (WML, 62fr [§15]).[7] This is, of course, another aspect of the *pharmakon*: even sublimation is pharmacological.

But the alternative to sublimation is what really exercises Stiegler. Sublimation may be ambivalent, absence of sublimation is pure poison. In psychoanalytic terms, the 'pulsions' or drives are bound together by desire and thereby transformed into libidinal energy (WML, 88fr [§23]), which can then be sublimated, or indeed released in *jouissance*. But the drives do not have to be bound: on the contrary, they can simply be expended in pleasure. The expenditure is far from benign, for it risks leading to the annihilation of what makes us human and indeed what makes our lives worth living. Contemporary consumerism and consumption is, for Stiegler, a truly pernicious side-effect of capitalism and of our technological progress, leading to an ever-increasing rapidity of satisfaction, a concomitant inability to sublimate, a loss of individuation and autonomy, and a disturbing uniformity and banalisation of discourses, sentiments and ultimately people themselves. Rather than yearning for the infinite, we are seduced by the triviality of mediocre and ephemeral satisfactions. This is part of what Stiegler calls 'proletarianisation' (WML, 88fr [§23]) and disenchantment (WML, 123fr [§40]), in which desire is destroyed to the extent that 'pulsions' are never allowed to remain unsatisfied (WML, 139fr [§45]) and libidinal energy is discharged rather than renewed (WML, 144fr [§48]). This in turn means that consumption replaces care, that is to say objects are destroyed rather than cherished; so not only are humans dehumanised, but the world and its goods are squandered. Stiegler's libidinal economy is also an ecology.

Desire, *jouissance, philia*

One of the best discussions of the dangers of contemporary consumerism comes in *Taking Care: Of Youth and Generations* (2008), where Stiegler lays out his view of the risks to future generations of unbridled and unmonitored technological advances. One of his primary *bêtes noires* is the addictive and cynical triviality of much television, which he sees as actively destroying the more humanising processes of individuation offered by the family and social integration. Children's television in particular often mocks the adult world, and with it the constructive values of shared laughter and tenderness which are replaced by a noxious complicity with irresponsibility and nihilism (TC, 8–9/24–5). This leads, Stiegler argues, to a deformation of the usual means of construction of a well-functioning transgenerational superego, paradoxically destroyed by an excess of censorship and a concomitant distortion of the psyche. What is happening, in the deceptive guise of fun and harmless pleasure, is in fact the liquidation of desire and the short-circuiting of experience in the constant drip-feed of short-term reward (TC, 12–13/30–1). I should perhaps say at this stage that Stiegler's analysis is, in my view, weakened by his puzzling failure to distinguish properly between pleasure and *jouissance*, which he comes close to conflating. Indeed he claims that the satisfaction of the pleasure principle, unmoderated by the reality principle, is what produces *jouissance* (TC, 5/18), and that *jouissance*, unlike desire, has the structure of 'jetabilité', which is to say it is disposable, made to be thrown away. Lacan would certainly not recognise his own understanding of *jouissance*, with its dark and powerful forces, in this picture of a mere escape valve for pleasure. On the contrary, Lacan *opposes* pleasure and *jouissance*: the former is what Eros and the other 'pulsions' all seek, but it is in fact a trap and diversion. Pleasure, even that of orgasm, usually termed *jouissance* in French (hence many of the apparent paradoxes), puts an end to desire but satisfies without slaking. Physical desire (or need) may be temporarily muted, but true desire for the Other, which arises from the unconscious, remains as acute as ever. This is why the pleasure principle is not really the life force we would expect, but seems to have a darker side and be linked obscurely to the death drive. Pleasure keeps us away from the depths of our desire ('the pleasure principle, which can only have one meaning – not too much *jouissance*', Lacan 2006b: 108), and this helps life to run more smoothly without excessive pain, but also masks from us the way in which desire is linked at heart to death; the subject may lose conscious control at the point of orgasm ('l'évanouissement du sujet', Lacan 2006a: 774/653), but true *jouissance* would entail death.[8]

Nonetheless, my possibly pedantic objection to his conflation of pleasure and *jouissance* in no way undermines Stiegler's main argument, which is that the short-circuiting of desire by premature and easy satisfaction interferes with both the processes of subjectivation (TC, 166/296) and the projection of a future for both individual and society. What is more, it leads to a highly undesirable social homogeneity, quite the opposite of the singularity and individuality produced by 'primary psychic and collective identification', which engenders not identity but rather alterity (TC, 61/114). Sharing the short-term pleasure of mindless TV, Stiegler argues, creates more destructive similarities of aim and experience than sharing long-term goals in the creative process of transindividuation. Libidinal economy needs to be saved from its current risk of dissolution in the potentially disastrous liquidation of primordial narcissism. In a sense the *pharmakon* of technics has gone from being the remedy that enables humanity to overcome the *défaut d'origine* that lies at its heart to the poison that destroys individuation and replaces it with mere subsistence.

Stiegler's own diagnosis of this situation is of course political as well as psychoanalytic: he maintains that the exploitation of desire is part of the second cycle of capitalism, after the exploitation of labour. To discuss this in any detail would go far beyond the scope of this essay on love and desire. Nonetheless, Stiegler's overriding position is clear: there is an urgent need for a new libidinal economy which will combat negative sublimation and desublimation and enable us to stop short-circuiting the processes of transindividuation. To use his terms, we must reinstil *philia* in the *polis* – for without *philia*, claims Aristotle, no city is possible (see Nagle 2005: 198) – and start to cherish again our desire for the infinite, even in a world without God. Stiegler has a striking image of our current predicament which he uses against himself: that of his smartphone, which he compares to the shell of a snail. It is a portable computer, via which he can retrieve all his own texts at any moment, to read or listen to. Stiegler describes this as a form of autoerotic and pathological narcissism with which he increasingly identifies. It is a *pharmakon* where the remedy of convenience and immediacy is clearly counter-balanced by the poison of hypertextualisation and hypermediatisation (Stiegler 2009: 34). Stiegler takes over from Sylvain Auroux the term 'grammatisation' to describe the history of the exteriorisation of memory and its (negative) consequences in the ensuing deprivation and weakening of knowledge in proletarianisation (Auroux 1994).

What then can we do to recover *philia*, care, attention, love and desire from proletarianisation? Surely the consequences of rapid technological progress cannot really be beyond our control. Stiegler's answer would seem to lie in philosophy as Plato understands it: philosophy as therapy for the soul, which is simultaneously, he says, also psychotherapy, sociotherapy and even care for the body (Stiegler 2009: 34). Philosophy is necessary in so far as it opens the way for an understanding of politics as the organisation of desire and 'a sublime form of transgression' (DD2, 31/54), and may 'become the basis of a renewed political project of philosophy where the main stakes are in technics' (AH). Our new technological milieu requires 'the constitution of a new milieu of psychic and collective individuation', a new social organisation, and a new libidinal economy without which no future aims and purposes will be possible if we accept that there can be no *telos* without desire. Philosophy is the key that will enable us to 'rethink technology' (Stiegler 2009: 35) and thereby recover what we have progressively lost from sight: our human 'défaut d'être' and all its consequences.

Conclusion

We have seen that for Stiegler technics is the unthought, the repressed: we enjoy its fruits and are in turn consumed by it quite unawares. Once we start to *think* it, we will have a chance of changing the balance of good and bad in its pharmacology: we will perhaps even be freed from ever more rapid *pulsions* and the ever decreasing returns of almost pointless pleasure, restored to love, desire, *philia* and sublimation. The restoration of our autonomy requires us to recognise our human, all too human, dependence on our prostheses. Technics may then once again enhance us rather than cause us harm in an inevitable *return of the repressed*, all the more noxious because unacknowledged and denied. And if we are tempted to think that Stiegler is an unreconstructed romantic optimist, or even a conservative traditionalist, we should perhaps bear in mind that he is not as deceived as we may be by the positive connotations of all the processes that he is recommending. Desire, sublimation and love may be infinitely preferable to *pulsions*, pleasure and consumption, but Stiegler is not nostalgic: he knows we have no origin to return to, that sublimation is itself a *pharmakon*, and that love is, ultimately, merely a (good) pathology, and the highest form of addiction (DD3, 82, 116).

Notes

1. Cited by Stiegler in DD1, frontispiece.
2. Sartre 1960: 131; Foucault 2001a: xxv/15; Lacan 2006b: 108; TT1, 136/146.
3. For an interesting discussion of this, see Gerald Moore's 'Crises of Derrida: Theodicy, Sacrifice and (Post-)deconstruction' (Moore 2012).
4. This text has unfortunately been given an inaccurate and misleading English title: *To Love, To Love Me, To Love Us*. A more appropriate translation would be: *Loving, Loving Ourselves, Loving One Another*.
5. The term 'primordial narcissism' was used previously by psychiatrist Ernst Simmel in 1944 to refer to an early stage in the infant's development before it is aware of its individuation from others.
6. See Laplanche et Pontalis, *Vocabulaire de la Psychanalyse* (1973: 338/261) for a clear account of the complexities of the notion of narcissism.
7. And desire itself is intrinsically a 'puissance de sublimation' (CE2, 17).
8. For a more detailed discussion of this aspect of Lacan, see Howells 2011: 131–47.

9

The Technical Object of Psychoanalysis

Tania Espinoza

Ressentiment and denegation are factors of ruin, as well as irreducible tendencies, which Nietzsche and Freud placed at the heart of their reflections a century ago. They will never have been exemplified so diversely as today. The reader will know then, that these authors, if seldom quoted in these pages, form the vanishing point of the perspectives I have attempted to open.

Bernard Stiegler, *Technics and Time, 1: The Fault of Epimetheus*[1]

In the late 1980s, the British psychoanalyst Christopher Bollas coined the term 'unthought known' to describe the unconscious (Bollas 1987). If 'technics is the unthought of Western thought', as Bernard Stiegler boldly states from the beginning of his oeuvre, then to think technics is also to think psychoanalytically about thought (TT1, 6). From this premise, two claims about the relationship between Bernard Stiegler's philosophy and psychoanalysis can be made. First, that technics is the unconscious of psychoanalysis. Second, that psychoanalytic technique, in so far as it looks at the underside of discourse, is central to the philosophy of technics. Yet Bollas's double concept adds a third claim, namely that psychoanalysis *knows* something about technics, even if this knowledge needs to be made manifest. The place to look for this unconscious knowledge, from the perspective of Stiegler's philosophy, is fairly obvious: one must turn to the technical object of and in psychoanalysis: the tool, the toy, the machine, those prostheses disseminated throughout the theory, the myths and the clinical cases. These, according to Stiegler, supplement the human in its lack of essence.

Interest in the psychoanalytic supplement or artefact has a rich and varied history, only a fragment of which can be revisited in this chapter. An example Stiegler is particularly fond of is

Paul-Laurent Assoun's reading of Sigmund Freud's *Totem and Taboo* in 'The Freudian Arsenal', which highlights the appearance of 'a new weapon' in the killing of the father of the primal horde. This product of technical invention constitutes an important 'contingent and material mark' in the mythical account of the origin of culture (Assoun 1990: 52, 61). Yet the crucial reference for Stiegler regarding this subject is the work of Donald W. Winnicott and in particular his concept of the transitional object. It is by reconstructing the dialogue between Freud, Winnicott and Stiegler that psychoanalysis's 'unthought known' of technics might emerge, through the figure of the object, as a way of elucidating Stiegler's recent discussion of pharmacology.

The pharmakon as transitional object

The first *pharmakon* is, according to Bernard Stiegler, the child's first possession, or 'transitional object' (WML, 13fr [Introduction]). Psychoanalysis thus allows Stiegler to bring together the child's chronological beginning and the condition of possibility of the pharmacological – that is to say, technical – life of an individual. This object, which negotiates the differentiation between subjectivity and objectivity, is said to owe its place in psychoanalysis to clinical observation, and is attributed a distinct reality within the child's early development: it appears, says Winnicott, 'at about four to six to eight to twelve months' (Winnicott 2005: 6). Therefore, unlike Freud's primitive horde of brothers, or Plato's Egyptian tale of the origin of writing in the *Phaedrus*, Stiegler's primal scene of pharmacology is not projected onto a mythical past, nor are there any warnings, as in Jacques Lacan's mirror stage, that its position in a diachrony is deceptive, even if it allows for 'wide variations' (2005: 6). According to Stiegler, however, the object does not so much *exist* as *consist* within the child's history.

In Stiegler's work, 'consistence' applies to objects that, by being infinite, open onto a world of idealities where the development of the spirit – which makes life worth the pain of living – becomes possible. It is in this sense that Stiegler speaks of the *pharmakon* as the origin of the life of the mind/spirit (WML, 15fr [Introduction]). If Stiegler's rewriting of the psychoanalytic account lies precisely in this substitution, that is, in elevating the transitional object to the ideal status of an object which 'does not exist but consists' (WML, 12fr [Introduction]), then a zone of indetermination between philosophy and psychoanalysis opens up. This zone is the domain of 'illusion', equivalent to Winnicott's 'potential space', which designates a beyond as much as an in-between. The *pharmakon* brings us to the beyond of the 'Analytic of Principles' in Kant's *Critique of Pure Reason* (1781), where the categories of thought no longer correspond to an object in reality, but rather to objects (God, the immortality of the soul, the ideal of cosmopolitanism) to be constructed by practical reason. It brings us also to a beyond of the pleasure principle in Freud's metapsychology, where desire is no longer ruled by a mechanistic economy of decreasing excitation; and finally, to a beyond of the *automaton* in Lacan's *Four Fundamental Concepts of Psychoanalysis*, where chance, or *tuchē*, breaks the repetitive motion of the drive to produce an encounter with the real. The domain of transitional phenomena designates in Stiegler's work precisely that region in between the dead, inert mechanical being and the living, organic biological being, where the technical being belongs and which is 'constitutive of libido' (DSL).

As 'organised inorganic matter' (TT1, 163–4), the technical object functions as a hinge between the tendencies and counter-tendencies that structure Stiegler's thinking, such as diachronisation and synchronisation, individuation and disindividuation, therapeutics and toxicity, sublimation and drive. These tendencies incessantly 'compose' with each other, or are in a tension that he calls metastability (WML, 196–7fr [§67]). They constitute the positive and negative poles of the *pharmakon*, the transitional object *par excellence*, both a potential cure and a potential poison. Stiegler's beyond–in-between is therefore a technical milieu that allows the relationship between the inside and the outside of a system – be it a cell, an individual, a group – or between what palaeoanthropologist André Leroi-Gourhan calls 'interior or ethnic milieus' and 'exterior milieus' (WML, 188fr [§65]). This relationship, according to Leroi-Gourhan, is constituted by the process of internalisation and externalisation of technical objects. Similarly to how Freud describes the *ego* in *Beyond the Pleasure Principle*, these objects are imagined as a membrane between interior and exterior (1955: 27).

For Stiegler, there is no interiority prior to the internalisation of an object, or a 'transitional exteriority' (WML, 41fr [§9]). However, there are latent, universal technical tendencies or potentialities that come from the interior milieu and that encounter a limit in the exterior milieu, becoming thus concretised and differentiated (WML, 188–91fr [§65]). These tendencies are what Stiegler has more recently conceptualised as 'traumatypes' and reproached Freud for not seeing when, in *Beyond the Pleasure Principle*, the latter postulates trauma as an excess of excitation coming purely from the outside, leaving the fragile ego membrane unable to protect itself. Stiegler retorts:

> Now, the organism cannot be affected by an exterior traumatism except when it is expected, except when, being protentially charged, it is touchable, affectable by this exterior traumatism that is already within it, and that is thus not totally exterior. Otherwise, either it would not be affected by it, or it would be simply destroyed. (DSL)

In other words, Stiegler does not find, in Freud, the theorisation of a primary inscription of the outside in the inside that would make the latter affectable. This returns us to the 'primal scene' of pharmacology as the site of that originary relationship between subject and object, which Stiegler will find in Winnicott.

Inscribable potential space, or the technical cultural milieu

Winnicott's paper 'Transitional Objects and Transitional Phenomena', presented at the British Psycho-Analytical Society in 1951, and developed in *Playing and Reality* in 1971, attempts to explain the way that a child establishes a successful relationship with the outside world, which, in adult life, will determine his participation in and contribution to culture. The possibility of such a relationship presupposes a distinction between what is interior (fantasy) and what is exterior (reality) to the child, yet happens in an intermediary area of play that is both interior and exterior, or 'extimate' in the vocabulary of Jacques Lacan. Ideally, it allows the transition from a state of illusory omnipotence where all the baby's needs are fulfilled by a devoted mother figure to one in which frustration is accepted. This passage explains, as it were, the transition from the pleasure principle, seeking to attend to the demands of the ego, to the reality principle, seeking to comply with the demands of society (Winnicott 2005: 13).

Winnicott's hypothesis is that such a space is embodied in an object that he therefore calls transitional. This object, a teddy bear, a favourite blanket, a piece of cloth to which the baby becomes exceptionally attached, is a substitute for the mother's breast – itself already a symbol for the maternal function of care – which needs to be present in order to produce the coincidence between the object that is invented because wished for by the child and the real external object that is discovered. A 'good enough mother' is one who is able initially to fulfill the baby's physical and emotional needs, thereby producing harmony between the child's imagination and reality. She becomes the support of transitional phenomena, which give the child the experience of creativity. For Winnicott this experience is ultimately what makes 'the individual feel that life is worth living' (2005: 87), a phrase that inspires Stiegler's *Ce qui fait que la vie vaut la peine d'être vécue: De la pharmacologie* (*What Makes Life Worth [the Pain] of Living*, 2010).

The theoretical default that sets Winnicott on his exploration of transitional phenomena is what he perceives as a missing place in the Freudian topology: 'the location of cultural experience' (Winnicott 2005: 128–9). *Playing and Reality* and much of Winnicott's theoretical and clinical work are part of an attempt to carve this space, or to answer the question 'Where are we when we are [. . .] enjoying ourselves?' (2005: 142). Creative apperception or play, according to Winnicott, takes place in the 'potential space' between the mother and the child, where reality is negotiated – but never definitively – between the not-me of the mother's body and the me of the child's body. This is a space created by trust and reliability, which depends on the consistency of care, symbolised by the continuity of the mother's presence for the child. An undeveloped system of retentions allows for the child to be separated from the mother for a short amount of time without this continuity being broken, yet, when the time of separation exceeds the child's memory of the mother a break occurs, leading to trauma. From the point of view of the child, absence is not understood as anything other than death (2005: 29). 'By potential space', says Winnicott, 'I refer to the hypothetical area that exists (but cannot exist) between the baby and the object (mother or part of mother) during the phase of the repudiation of the object as not-me, that is, at the end of being merged with the object' (2005: 144). Accepting the paradox regarding existence is key to the understanding of transitional space, for what Stiegler calls the 'retentional apparatus' is formed through this fluctuation between being and not-being-there.

Very much like Kant's noumenon, which is the object of the 'Dialectic of Illusion' in the *Critique of Pure Reason*, this space between the mother and child is an empty place that needs to be 'filled', and is best symbolised by the string that separates by uniting, an image to which we will return. The fragile matrix of relatedness and care constitutes the room for 'play', which is the first instantiation of cultural experience. The reason for this is that invention and discovery overlap and neither psychic reality (hallucination) nor physical reality has priority over the other. This space is thus marked by the possibility of variation, in two different senses. On the one hand, reality appears to the child as malleable. As long as continuity is preserved there is no experience of something falling completely outside of the child's control, nor, due to the fallibility of the mother – who therefore could not be replaced by a finite, predictable machine – is there an experience of absolute omnipotence of thought, for there is a margin of failure on the part of the carer. Potential space therefore lacks the fixity of inner reality (which, unmediated by an other, becomes autistic) and of the external world (which we might perceive as 'ready made'). For Stiegler, this malleability makes the space pharmacological, allowing both for a relationship to the object of adoption, which remains creative, or a relationship of adaptation, where the fixity is not challenged (WML, 42fr [§9]).

On the other hand, this experience is, necessarily, unique to the individual and thus allows for infinite variation: 'there is a kind of variability here', says Winnicott, 'that is different in quality from the variabilities that belong to the phenomenon or inner personal psychic reality and to internal or shared reality' (2005: 144). Such variability is compatible with the universality of the phenomenon, just as particular 'ethnic milieus' can be the result of universal 'technical tendencies'. The dissonance between the two makes this a space of individuation, understood in the language of Gilbert Simondon as the 'de-phasing of the individual with respect to her preindividual milieu' or external reality' (WML, 55–6fr [§13]). Winnicott seems to conceive this variation as a result of transitional phenomena being part of ontogenesis (individual development), for potential space marks the 'continuity with personal beginning', as opposed to phylogenesis (development of the species). We see this, for example, in his claim that 'the special feature of this place where play and cultural experience have a position is that it *depends for its existence on living experiences,* not on inherited tendencies' (2005: 146). Yet while he does insist on carefully differentiating these phenomena from those having 'instinctual backing' (2005: 132) and thus belonging to phylogenesis, in so far as 'living experiences' are mediated by the transitional object – by tertiary retentions, in Stiegler's vocabulary – they are already epiphylogenetic. Another way of looking at this would be to say that transitional phenomena form the bridge between ontogenesis and epiphylogenesis, understood here as the development of what Winnicott calls culture:

> In using the word 'culture' I am thinking of the inherited tradition. I am thinking of something that is in the common pool of humanity, into which individuals and groups of people may contribute, and from which we may all draw if *we have somewhere to put what we find*.
>
> There is a dependence here on some kind of recording method. (2005: 133)

This formulation shows the equivalence between the 'somewhere to put what we find' and a 'recording method'. The potential space of culture is not only one that needs to exist so that it can be filled, but it has to be one where something can be inscribed; it is a technical milieu invested with tendencies and traumatypes. However, Winnicott, unlike Stiegler, wants to maintain the order of priority and appearance between the potential space of trust that is unique to an individual carer–child relationship, and the potential space of cultural heritage that is 'in the common pool of humanity'; where the former must come before and is the condition for the latter (2005: 148). This, to some extent, produces a deterministic account of individual development where someone who lacked this primordial care is condemned to exist, or subsist, outside of cultural experience – an account that contradicts Winnicott's own biography as the son of a depressive mother.

From the point of view of Stiegler's pharmacology, we could propose two observations. One would be to point out, very simply, that the matrix of care that Winnicott takes as a natural extension of the mother (almost as the result of an 'inherited tendency' that he otherwise rejects) is itself dependent on 'inherited tradition' and thus on certain mechanisms of cultural transmission capable of producing the 'maternal function'. If the technicity of these mechanisms was concealed from Winnicott, it is apparent to us perhaps because their effectiveness can no longer be taken for granted. The anxiety that our society may structurally destroy the 'good enough mother' runs throughout Stiegler's *Taking Care of Youth and Generations* (2008). However, this 'good enough mother', for Stiegler, is never simply an individual, but is made up of a functioning state, a 'good enough state' capable of regulating the predatory tendencies of the market as well as of a cultural milieu where 'consistencies' or spiritual objects are accessible to all in a public domain.

The second observation has to do with the therapeutical power of both psychoanalysis and culture. Winnicott's conception of patient–analyst transference as a repetition of the potential space between mother and child is already the beginning of a possible refutation of the determinism mentioned earlier. By technically recreating the conditions of play, psychoanalytic practice would be one of the cultural developments that allows us to repair the eventual breaks in the necessary continuity between the individual and the parental figure. Thus, recognising the primordial technicity of the potential space makes it easier to acknowledge the possible therapeutical use of psychoanalysis with respect to apparently irreversible states, reversibility being one of the properties of our pharmacological condition.

To think pharmacologically, within psychoanalysis, would be to shift the focus of concern from the death of the father and the dissolution of the symbolic order – events that for Stiegler would constitute changes in the technical *epokhē* – to the possibility of destruction of the maternal 'potential space', the more primordial condition of possibility of creating or adopting a new order or *epokhē* therapeutically. The resurrection of God, as the thoroughly material infinity of desire, which seems to be one of the philosophical tasks of a pharmacology of the spirit, necessarily passes through the recuperation and preservation of the technical milieu of illusion, the location of cultural experience.

Fiction and the infinite object: a cotton reel

If the substance of transitional phenomena is illusion, we could now describe the process that Stiegler calls infinitisation – the necessary step from existence into consistence – as a sort of fictionalisation, through the transitional object of the psychoanalytic case study. The relation between fictionalisation and transitional objects is found in a well-known and emblematic story: Freud's grandson's *fort–da* game, described in *Beyond the Pleasure Principle*. Pausing abruptly in his discussion of 'the dark and dismal subject' of the war-induced traumatic neuroses that occupy him for the most part, Freud decides to turn to 'one of the earliest *normal* activities' of the mental apparatus, namely 'children's play' (Freud 1955b: 14). He recounts the story of an eighteen-month-old child who, having acquired the habit of throwing objects away, invents a peculiar game that consists in making a cotton reel disappear, then reappear. The throwing-away motion is accompanied by the sound of 'o-o-o-oh', which Freud interprets as the German word for 'gone' ('*fort*'), while '*da*', meaning 'here', accompanies its return (1955b: 14–15). We also learn that the child's mother, Freud's favourite daughter Sophie, happens to absent herself quite often from the household. Freud interprets the game as the child's response to these involuntary exits and its wish to master them.

Winnicott's reading of the *fort–da* game precedes his paper on transitional objects by a decade. The cotton reel only appears in the latter through its resemblance to Winnicott's case study of the 'string boy' (2005: 21–7) – thus stylised and reworked into another fiction. Yet Winnicott's relationship to this episode in Freud's writing, is, from the outset, one of incorporation. Winnicott reinterprets the game by giving an entirely positive reading of it as 'regaining happiness' for the child, in a way that opens the field for his theorisation of transitional phenomena (Winnicott 2011: 52). In order to do this, he has to translate it away from the ambivalent and death-ridden context of *Beyond the Pleasure Principle*, into his 'large' room in the hospital, where he has been observing babies of between five to thirteen months of age in what he calls a 'Set Situation' (2011: 37). The situation consists in placing a shiny tongue depressor or right-angled spatula on the side of Winnicott's desk, within the reach of the child who has been brought to consultation, and who is sitting on his mother's lap. Generally, after a moment of hesitation, the baby reaches for the spatula and puts it into his mouth, to then play at throwing it repeatedly on the floor. This 'incorporation, retention and riddance' constitutes the 'normal sequence of events' (2011: 51).

The clinical interest in the situation lies in the way that the child's opportunity to establish a connection to the object, to play with the spatula, has an immediate therapeutical effect. One child ceases to have attacks of asthma; another overcomes her 'feeding disturbance' and fits of anxiety (2011: 41–3, 39). An explanation of the cure seems to lie in the reconstitution of the circuit of desire in a safe, stable environment where the 'full course of an experience' can take place (2011: 50). The child is allowed to pass from a moment of hesitation or mental conflict – where the fantasy of eliciting anger in a strict mother through his action competes with his interest in the spatula, producing anxiety – into one of 'acceptance of the reality of desire' (2011: 38) and consequent joy in handling the spatula. Finally, the moment of 'riddance' is what allows the child to pass from desiring one object to desiring another, launching the infinite metonymy of desire.

The philosophical interest, on the other hand, from a Stieglerian point of view, lies in the fact that this 'Set Situation' can be said to recreate the primordial relationship of the human to the technical object, establishing it as necessary for relating to the external world and others, but also to the child's own body, while at the same time suspending its naturalness. To place the cure as a reparation in that initial gesture of 'grabbing' the tool – for the spatula that Winnicott offers to the baby is, in the first place, a medical tool for examination – suggests that this gesture is thoroughly pharmacological. The artificiality of the 'Set', its status as 'instrument' of observation, emphasises the fact that relatedness is technical from the outset. Furthermore, Winnicott's analysis reinterprets the medical tool – one invasively and often forcibly put in the mouth of the child – as one that can restore health by being played with, where the condition of play is that the baby himself takes the initiative to bring the object into his mouth as a proto-autonomous recognition of the desire on which he acts. The tongue depressor, very much like bodily organs in the course of evolution, is in this process defunctionalised and refunctionalised. Winnicott's 'Set Situation' is thus as much a therapy for the children as for the clinical practice of paediatrics. As Freud had anticipated in explaining the analogy between children's games and dreams, 'If the doctor looks down a child's throat [. . .] we may be quite sure that these frightening experiences will be the subject of the next game' (1955: 17).

Winnicott identifies the *fort–da* game as the third stage of the larger sequence of moments in relating to objects that he had identified, namely 'riddance', and thus as one which 'implies' the two preceding moments. Yet it is only through the reference to the cotton reel and to Freud's theorisation that Winnicott can complete the analysis of this stage. He is able to incorporate Freud's successive readings of the meaning of the game: as mastering the mother's absence, but also as an act of vengeance for the mother's abandonment, by pointing out that there are, in fact, two mothers at play. There is an external mother and the 'counterpart in his [the child's] mind' or an internal mother (2011: 51). The game's effectiveness, according to Winnicott, lies in allowing the representation of the internal mother. The child, says Winnicott,

> externalises an internal mother whose loss is feared, so as to demonstrate to himself that this internal mother, now represented through the toy on the floor, has not vanished from his inner world, has not been destroyed by the act of incorporation, is still friendly and willing to be played with. (2011: 52)

What seems to be missing from this analysis is that the very 'internalisation' of the external mother has also been mediated by the object in the stage of 'incorporation'. Winnicott's doubling of the mother into an internal and an external one comes from the recognition that internal and external reality, as Stiegler points out, cannot be opposed (DSL). If what Freud is aiming at, in *Beyond the Pleasure Principle,* is a theory of that transition and passage between the two, Winnicott realises that the cotton reel is what allows this passage.

The mortifying, revivifying object of desire

One of the most important arguments of the philosophy of technics is that hominisation is inextricably linked to the prolongation of living memory – the brain – through dead memory – the objects of tertiary retention. Technics is 'the continuation of life by means other than life' and evolution can only be understood as a dialogue between living and dead memory, which allows transmission (TT1, 17). The object created through play is, in our psychoanalytic example, a nexus between the dead and the living, yet the distribution of life and death complicates the one presented above. It is not only that the cotton reel is dead, as opposed to the child who is alive. It is the mother who is both alive and dead in the psyche of the child (and of Freud, who writes the episode before the death of his daughter Sophie and finishes it afterwards). The potential loss of the internal mother is already a form of death, which representation and alienation in the object, paradoxically, counteracts:

'Something is missing until the child feels that by his activities in play he has made reparation and *revived* the people whose loss he fears' (Winnicott 2011: 52 [emphasis added]). The cotton reel is able to 'revive' the mother, because the representation separates the internal from the external mother so that the internal one can survive the latter, when she leaves. The transitional object, therefore, is neither dead nor alive but, as a good *pharmakon*, can bring about both revivification (Winnicott) and mortification (as Lacan understands the subject's alienation in the object).

In his Seminar XI, Lacan models *objet a* precisely on the Freudian toy:

> If it is true that the signifier is the first mark of the subject, how can we fail to recognise here – from the very fact that this game is accompanied by one of the first oppositions to appear – that it is in the object to which the opposition is applied in act, the reel, that we must designate the subject? To this object we will later give the name it bears in the Lacanian algebra – the *petit a*. (Lacan 1978: 62)

The cotton reel is, in Lacan's text, the transitional object that belongs both to the object (*automaton*) and the subject (*tuchē*, meaning 'chance'), following the Aristotelian distinction between these terms as applying to non-intentional and intentional beings, respectively. It is also what makes the two coincide: the subject is, according to Lacan, in the object. Returning to its fictionalised appearance in *Beyond the Pleasure Principle*, the cotton reel allows Freud to move from the repetition that can still be explained within the paradigm of the pleasure principle – as in the multiple interpretations of a staging of the disappearance of the mother and her return, as a consequence of a desire for mastery of this contingent event in the life of the child, and a desire for revenge – to a beyond. This beyond constitutes *tuchē* as an 'encounter with the real' that is produced, for Freud, by the cotton reel. It is through this object that the text introduces, in a footnote, the death of Freud's daughter Sophie, Ernst's mother. Lacan exemplifies this encounter with the dream of the burning child in which a father confronts the unspeakable death of his son in a dream (1978: 59). The *fort–da* game is Freud's restaging of this dream for the death of his daughter.

Here Lacan places the technicity of the object in its being able to signify an opposition, that is, in its symbolic nature. It is to this extent that Lacan can claim, earlier in the passage that 'man thinks with his object' (1978: 62). This ability of the object is, however, itself grounded in its material/technical constitution; its being made of wood and string and therefore able to be thrown (*fort*) and brought back (*da*). In Stiegler's words, this relationship could be described as follows: 'The pharmacology of spirit is a pharmacology of *symbolic* relations, but such that objects are the first instances of these relations, and insofar as that which the Greeks call a *sumbolon* is an object' (WML 117fr [§38]). Stiegler continues:

> Before constituting themselves hypomnesically, the circuits of transindividuation through which the spirit is formed are deployed on the basis of infantile transitional relations, and as objects *invested with spirit* in the sense that Husserl will use this expression to designate the book, yet enlarging its domain to all everyday objects. (WML, 117fr [§38])

Winnicott's spatula, existing within the symbolic universe of a transgenerational dialogue in psychoanalytic theory, is never a mere physical object, but one invested with the spirit of the mother who leaves too soon, or the doctor 'looking down a child's throat', whom Freud anticipates will be 'the subject of the next game' (as well as the subject of an infamous passage in Freud's *Interpretation of Dreams* (1900), the 'Dream of Irma's Injection'). These fictional ghosts constitute the spatula's materiality as much as its angular shape or shininess. They make for the objects' attractiveness and our desire to incorporate them, retain them and finally detach ourselves from them, though not without having been transformed in the process. The technical object is thus also a spiritual object, implicated in and constitutive of a technical unconscious, constantly threatening to dissolve the fragile 'interior milieus' that still constitute subjects' singularity. Threatens, that is, unless we learn how to play. For, if games are about 'frightening experiences' (1955: 17), they also – and this is the lesson from Winnicott – allow for their reparation. The technical reconstitution of the never-good-enough–good-enough mother is one of the possibilities that Stiegler's philosophy offers psychoanalysis of redeeming and letting go of a figure that has been described as the eternal referent of the 'unthought known' (Rose 1996: 111). A philosophy of technics constitutes perhaps that third stage of psychoanalysis's relationship to the object – riddance – in which the mother 'is still friendly and willing to be played with', while the parental function is found and critiqued elsewhere, in the technical milieu, the potential space of culture in relentless need of care.

Note

1. TT1, x

10

Desublimation in Education for Democracy
Oliver Davis

'Philosophy – as deambulation, dialogue and writing – was to the Greek world what shooting-up is to the post-Fordist West.'

Beatriz Preciado, Testo junkie[1]

In this desublimatory quip it is not quite clear what Beatriz Preciado means by likening philosophy's function in the ancient Greek world to shooting-up in the post-Fordist West. What *is* clear is that we are talking here about 'pharmacology' in a desublimated sense far removed from Stiegler's elaborate – highly sublimated – refiguration of that term; clear too is that Preciado's vision of philosophy and shooting-up as analogously interruptive of the cycle of a society's necessary self-reproductive work could hardly be further removed from Stiegler's . . . and doubtless on both points. It is not quite clear what Preciado means (nor indeed to what extent she really means it: is she merely likening two socially stigmatised practices, or going further to posit their equivalence?), just as it is never quite clear what exactly is going on whenever sublimation and desublimation enter the theoretical picture. As Craig Saper speculates, suitably inconclusively: 'Perhaps sublimation itself clouds the issues and confuses desires' (Saper 2009: 65). Clouds of smoke appear; with them a whiff less of sulphur than of alchemy, the subliming science of a master so deftly caricatured by his servant in Chaucer's Canon's Yeoman's Tale.

Sublimation very often enters the picture in Stiegler's work. Indeed, it is one among a nexus of central theoretical terms that appear to be tightly interconnected with a very high degree of systematic rigour in his uniquely compelling analysis of the interplay between technics, economics and culture in contemporary consumer capitalism. Powerful though this analysis is – neither its coherence nor its attunement to the contemporary moment are merely a matter of smoke and mirrors – my contention will never-theless be that Stiegler, like Marcuse before him, takes sublimation *far too seriously*. I shall endeavour to show that Stiegler's sublima-tory prescriptions for a new pedagogy, one dedicated to the culti-vation of 'consistencies', fail to give due attention to the political role that techniques of *desublimation* must also be allowed to play in properly democratic education. So I shall question the politics of the sublimatory paradigm Stiegler has adopted, in particular as it plays out in his prescriptions for new regimes of pedagogy and care-giving in *Taking Care of Youth and the Generations* and in the manifestos of Ars Industrialis. My reading is informed by, though scarcely refers directly to, Jacques Rancière's work on poli-tics and pedagogy. My suggestion will be that Stiegler's analysis can be deemed compelling not just by virtue of the persuasiveness of its theoretically elaborated – sublimated – formulations but also in the rhetorical strong-arm tactics he so skilfully deploys to *compel* acquiescence from his audience, strategies which are in large measure desublimatory and which he is not only reluctant to acknowledge, still less to theorise, but unwilling to envisage being shared on a democratic basis.

Let me first locate sublimation in the nexus of theoretical con-cepts at the heart of Stiegler's analysis. In his groundbreaking development of Marx's political economy, Stiegler argues that contemporary capitalism seeks to reverse the tendential fall in the rate of profit by extending the orbit of its exploitation from producers to consumers (CPE, 25–8). Consumer capitalism's purchase on its new victims is secured by way of 'psychotechno-logies' – prototypically television advertising – which seek to capture the attention of consumers and control that attention for their own ends, captivating their audience 'through the most archaic drives, then compelling it to construct a consciousness reduced to simple, reflex cerebral functions, which is always dis-enchanted and always "available"' (TC, 15). In controlling the attention of consumers, these psychotechnologies tend to destroy their capacity for sustained and 'deep' attention, above all when the audience in question is comprised of young people, whose developing brains are that much more malleable and therefore vul-nerable, although these psychotechnologies also conspire to effect in adults 'an unprecedented regression' to a state of 'adult immatu-rity' (TC, 23). What particularly concerns Stiegler in *Taking Care* is the way in which this direct appeal to 'the most archaic drives' is bound up with an undermining of the proper relationship between the generations, a relationship which implies the 'care' of children by adults, care in the context of which, he thinks, 'inherited symbolic representations, bequeathed by ancestors and transmitted by parents and other adults' are handed down from one generation to the next (TC, 8). The consequences of this breakdown Stiegler identifies as ranging from casual incivility[2] and delinquency (TC, 35) through riotous unrest (the book was first published some three years after the 2005 Paris riots) to murderous anti-social violence, as in the case of Richard Durn (RW, 65 [§6] and L) and 'total war' (TC, 86); not to mention 'addiction, cognitive overflow syndrome, attention deficit disorder, depression, impotence, and, finally, the collapse of desire' (TC, 42).

Stiegler understands capitalism as a form of 'libidinal economy', which is to say a system of harnessing and channelling unconscious drives, prototypically sexual drives. If capitalism were working well and had not overreached itself in the hyperindustrial consumerist turn just outlined, then this channelling would be for socially useful, humanising, individual and collective purposes. Following Freud, and more particularly Marcuse, Stiegler refers to such benificent channelling of the drives as 'sublimation':

> Libido is the socialisation of the energy produced by the sexual drive, but such that, as *desire*, this drive is trans-formed into an object *capable of being sublimated*: an object of love – the love of another way of existing, which is to say love as a *passionate attention* to the existence of another man or woman, or love as the *passionate and patient* attitude towards *consistencies*, which is to say *objects which do not exist yet which give to our existence those sublime forms of knowledge* by way of which *doing, living* and that very particular form of attention called *contemplation (theorein)* are *opened up* to us. (RW, 61 [§6])

'Consistencies' are sublimated – fragile – cultural objects of belief which include aesthetic and theoretical objects. The practice of tending to such objects is simultaneously humanising for the individual concerned, in a sublimation of archaic drives, and enriching of the social and material environment, by virtue of the rigorously interconnected circuitry which constitutes the 'whorl' of transindividuation (TC, 156). The sublimation of drives, which in their 'archaic' form demand instant satisfaction, into desire as a socialised and therefore deferred promise of satisfaction, is a prerequisite for access to those longer and intergenerational circuits of transindividuation which constitute the human spirit. 'Sublimation', Stiegler says in his book on aesthetics, is what enables 'participation in what Hölderlin called "the most high"', *Das Höchste* (SM2, 64fr [§15]).[3] If the danger of contemporary psychotechnologies of the internet age – running riot in an almost entirely unregulated and 'care-less' fashion – lies in their direct appeal, rather, to 'the most low', then the challenge Stiegler frames with singular perspicacity is how to tame their energies in a way which will allow them to be marshalled in support of the sublimatory and transindividuating purpose of the human, the human understood not as that which merely 'uses' (or indeed is determined by) technics, but as that which is coterminous with them:

> For every stage of grammatisation, societies institute therapeutic systems, systems of care, techniques of self and others, which constitute spiritualities and diverse noetic forms, from shamanistic models to artistic models, passing through churches, medical therapies, schools, sports, philosophies, and every system of sublimation. (CPE, 120)

My contention, however, will be that each and every therapeutic and care-giving system is more than merely a 'system of sublimation' that constitutes 'spiritualities' and 'noetic forms'; each also implies carefully organised techniques and political distributions of desublimatory power. For Stiegler, however, desublimation always seems to end up being toxic, both for the individual and – by destroying, or shortening, the circuits of transindividuation – for the social and material environment:

> Desublimation – which thus leads and in the same movement to the spread of negative externalities, to the liquidation of commerce by the market and to the destruction of social connectedness – can be translated by the fact that the bourgeoisie is swept away by the mafia, which is the fate befalling the former communist countries, but also of all drive-based capitalism. (CPE, 62)

I say desublimation always seems to 'end up' being toxic because this is perhaps less a matter of philosophical axiom – for the pharmacological status of his own work would stipulate, as a matter of axiom, that neither desublimation nor sublimation can be intrinsically toxic – than the way in which that axiom plays out in Stiegler's practice of writing, speaking and campaigning. Stiegler behaves as though the cure were only ever a matter of sublimation, never of desublimation, which is always, as far as I can judge, associated with the spread of 'negative externalities and pharmacological toxicity' (CPE, 49) into the milieu, and is therefore always understood as a poisoning which goes hand in hand with *'the degradation* [avilissement] *of the social and the destruction of all forms of intelligence'* (RW, 89 [§10]). However, to complete the picture on desublimation, as Stiegler understands it, I should mention that on occasion he also refers to something he calls 'negative sublimation' (DD2, 48, 98, 120–1), which seems in all essential respects to be the same as desublimation in its devastating consequences, but which specifically involves elaborate technical–organological complexes to magnify its toxic effects.[4] This concept of 'negative sublimation' seems to flicker in and out of his work in 2006, never to be heard of since.

Intermittence is Stiegler's Pascalian–Proustian–Aristotelian alternative to desublimation. Having freighted desublimation in the way I have just described, he needs another way of allowing for the fact that none of us can always have our attention focused on the higher things in life. Characterising the *amateur* of art objects, Stiegler writes that:

> he is one who *loves an object* and *sublimates* and thus believes in that object and who can sometimes lose 'faith', which is to say he no longer believes in his object. This love of the object, as a way of acceding to the noetic [*comme passage à l'acte noétique*], can be just as intermittent as participation in the divine. (SM2, 153fr [§33])

Echoing Pascal, Stiegler suggests that a condition of having sublimatory 'faith' in 'consistencies' is occasionally losing it; but rather than characterise such moments of failing in terms of desublimation Stiegler speaks instead, in Aristotelian vein, of the 'intermittence' of the 'noetic soul': 'the noetic soul is only actually noetic intermittently' and 'not only are all human beings potential artists but also [. . .] *no* human being is *ever* an artist *in actuality* other than intermittently' (SM2, 180fr [§38]). Stiegler goes to extraordinary lengths here to block out space that might otherwise be left for benign forms of desublimation. For even in these intermittent moments in which the *amateur*'s sublimatory 'faith' fails he will, as Pascal recommended to the believer, '*place his trust in the prostheses of faith that are the repetitions* in which consist what we call practices' (SM2, 153fr [§33]), even though presumably part of what losing such 'faith' means is precisely losing the ability to trust in those 'prostheses'. Not even a return to the wild (*ensauvagement*) or a fall from the heights of artistic grace are to be understood in terms of desublimation: both are rather to be conceived as particular forms of sublimation (SM2, 162fr [§35]) in a move which surely overgeneralises from the very particular 'arts of falling' indexed in the footnote's reference to judo and mountaineering. Fill in all the available theoretical space though Stiegler does – and this may be an effect of his not to be belittled, but invariably counterproductively anxious, desire to *take care* – Charcot's dictum, beloved of Freud, returns to mind: 'Theory is good but it doesn't prevent things from existing' (Freud 1977: 156). Sometimes a fall really is a fall, without being a disaster. Something in Stiegler's tightly-woven, catch-all theoretical frame can be let drop without this being care-less: his anti-democratic disavowal of desublimation. Desublimation must be allowed to happen. Proust – though not Stiegler's sublimatory Proust – realised this.

Stiegler's Proust is something of a two-dimensional nostalgic, an orchestrator of sublimation who harks back longingly to a lost aristocratic age of aesthetic participation: when Stiegler speaks of the tragic '*verdurinisation* of the world' (SM2, 148fr [§32]), he implies, by contrast, that for Proust and his narrator the Guermantes and the aristocratic order they represent were a touchstone of authentic, because participatory, aesthetic experience, whereas in fact the Guermantes are among the most ignorant people in the novel, in art as in other matters. The poetry with which their name is imbued is quite explicitly a trick of the narrative light. Whenever Proust is put to work by Stiegler (as in the epigraph at CPE, 71, for example), it is a Proust who has already been signed up to Stielger's own sublimatory agenda. Yet the real Proust will just as readily entertain the imperatives of desublimation and the complexity to which their cross-currents give rise; in this way he is far more akin to Freud, as Malcolm Bowie so acutely sensed:

Venice [. . .] provides Proust with a model not only of desire in perpetual displacement but of the sublimating and desublimating exchanges that occur between the sexual and the cultural realms. Architecture and painting divert and absorb the narrator's sexual impulses, but precariously. At any moment the gothic trefoil may drive him back upon his Oedipal longing, or a Carpaccio panel reawaken his passion for Albertine. (Bowie 1987: 93)

Freud also recognised that desublimation must sometimes be allowed to happen. Stiegler's own, arguably somewhat predictable, engagement with Freud consists principally in noting that he failed to understand the role of *hypomnemata* in the history of desire and sublimation (CPE, 41; SM2, 183–248fr [§§39–51]). Stiegler focuses excessively on the most biologistic and historical zones of Freud's account, where sublimation is apprehended over the course of human evolution in relation to the attainment of upright posture and the concomitant suppression of smell in favour of sight. Yet Freud was considerably more theoretically eloquent and therapeutically ambivalent about sublimation than Stiegler acknowledges. Moreover, by taking issue mainly with Marcuse's more straightforward pleading on behalf of a carefully orchestrated 'regression' of sexuality 'behind the attained level of civilised rationality' (Marcuse 1966: 161), Stiegler has chosen the path of lesser resistance when it comes to (de)sublimation.

'I have endeavoured to guard myself against the enthusiastic prejudice which holds that our civilisation is the most precious thing that we possess or could acquire' (Freud 1961: 144). Hans Loewald in his classic account reminds us that for Freud, both as theorist and more particularly as therapist, sublimation was 'at once privileged and suspect' (Loewald 1988: 1), and indeed seems to have been so since the concept first emerged in his writing in the 1890s. In his 'Recommendations to Physicians Practising Psycho-Analysis' Freud is careful to warn the doctor against pushing the patient to sublimate beyond his or her constitutional capacity:

It must further be borne in mind that many people fall ill precisely from an attempt to sublimate their instincts beyond the degree permitted by their organisation and that in those who have a capacity for sublimation the process usually takes place of itself as soon as their inhibitions have been overcome by analysis. (Freud 1958: 118–19)

One element in Freud's therapeutic thinking about sublimation which is left by the wayside in Stiegler's adoption of the concept is thus the sense that the doctor should exercise restraint and hold himself in check instead of foisting his own sublimatory agenda on those in his care. As Loewald comments:

> If the analyst is to be a civilising agent, in the best sense of the word *civilising*, then in *forcefully* directing the patient to higher aims he undoes his own work and contributes to further misery and 'discontent' instead of promoting the patient's own development. (Loewald 1988: 39)

Stiegler is obviously not obliged to take on every single element in Freud's account of sublimation, especially if Laplanche and Pontalis are correct that there is no 'clear and coherent' account of sublimation 'in psychoanalytic thought' (Laplanche and Pontalis 1973: 433/467). But the fact that he disregards Freud's warnings of caution must give us pause, not least because these warnings pertain most particularly to the therapeutic application of the concept, and Stiegler's interest is quite explicitly in social therapy. Richard Beardsworth is accordingly right to argue that, in Stiegler's highly selective reading of Freud on sublimation, he 'cuts the knot of this ambivalence too quickly' (see Beardsworth's article in this volume, p. 221 [2010: 195]).

This hasty and very partial assimilation may be all the more problematic because some recent scholarship has suggested that sublimation was considered by Freud himself as something of a joke, indeed a Jewish 'in-joke', invaluable in countering the accusations of pansexualism levelled against psychoanalysis and vital in securing cultural acceptance of it in a climate of anti-Semitism ... but not to be taken entirely seriously, or certainly not at face value, by those 'in the know'. In the discussion which followed a paper at a 1930 meeting of the Vienna Psychoanalytic Society, Freud is reported to have observed: 'People say, "This Freud is an abominable person; however, he has one rope, with the help of which he can pull himself out of the sewer [*Jauche*, literally 'liquid manure'] in which he dwells, and this is the concept of sublimation"' (Geller 2009: 170; Capps 2010: 267–8). Jay Geller and Donald Capps suggest, in their compelling accounts of the genesis and status of the concept of sublimation, that it was 'spun', indeed 'weaponised', almost from the outset; a smoke-bomb of sorts, a 'Trojan' vector for the psychoanalytic 'virus' embraced, as Capps notes, with somewhat simple-minded enthusiasm by the first wave of American psychoanalysts (Capps 2010: 284). Should this matter to Stiegler, who after all is not by vocation a historian of psychoanalysis? If sublimation were an incidental or ornamental presence in Stiegler's work then perhaps we could conclude that it should not. Yet sublimation, as I have indicated, is an integral part of a central nexus of key Stieglerian concepts. Indeed, in *Taking Care* he will assert that sublimation is the very engine of difference which powers the *pharmakon*: 'sublimation makes a "poison", for example, a *hypomnematon*, its remedy' (TC, 180). In what remains of this chapter I shall show that Stiegler's selective and incautiously enthusiastic adoption of a sublimatory agenda is of a piece with his failure to thematise the desublimatory insistence of his own discourse, and that this has disturbing political consequences for his philosophy of education.

In *Taking Care* Stiegler salutes the vision of statesman Jules Ferry, who made secular state primary education compulsory in France in the early 1880s, in the following terms: 'for Jules Ferry it was a matter of *substituting secular sublimation for religious sublimation*' (TC, 52/100) and, by making primary education a legal obligation, of instituting 'a positive power of sublimation – as a discipline of transindividuation that in turn fosters the political maturity that is the essence of *Aufklärung*' (TC, 126–7/227).[5] Michel Foucault, in his account of disciplinary power and its institutions in *Discipline and Punish*, is accused by Stiegler of having failed to acknowledge this 'positive power of sublimation' and in particular of having failed to understand that the Ferry laws were democratising 'technological' instruments, in that they made it clear that the State's expectation of all of its children was that they accede to literacy and thereby to political maturity. Foucault failed to see the Third Republic's school system for what it was: a system of care designed to form deep attention on an industrial scale and thereby to make France a mature democracy.

Foucault is not alone, Stiegler suggests, in his one-sided view of Republican schooling; his *ingratitude* and indeed his ignorance is felt to be typical of an overprivileged generation of theorists associated by Stiegler with May '68. It is beyond the scope of this chapter to examine in detail quite how Stiegler 'violently unravels the Foucauldian fabric' (Turner 2010: 255).[6] What I want to remark on instead is the untenable paradox that this defence, by Stiegler, of what he calls a 'positive power of sublimation' proceeds by way of an unravelling – the sublimated content of which is the predictable charge that Foucault fails to understand the impact of grammatisation and *hypomnemata* – that is, at the same time, a violently desublimatory *coup de force*. Moreover, it draws strength from a reactionary natalist rhetoric, indicative at the very least of an extreme form of 'reproductive futurism' which is somewhat more visible in the French than the translation.[7] For at issue is Foucault's 'fecundity' (*fécondité*), his 'fruitfulness' and 'fertility', and indeed that of the '68 generation with which Stiegler is all too ready to conflate him; in question is whether their methods and tropes are 'fertile' (*féconde*) (TC, 121/219). Foucault's account of the school as an institution of disciplinary power is typical, Stiegler argues, of what Stephen Barker, the translator, tells us is a '*disappointment* mind-set':

> This *disappointment* mind-set, evident in many famous figures newly graduated from, for example, the ENS in France, and from top American universities and colleges – particularly after the hugely disappointing historicopolitical abortion called '1968' – has transformed its performative power from a discourse of *initiation* [*déniaisement*] based on the rhetoric of *only, ne ... que ...* [as in 'it was *only* a dream'] into the future anterior: 'school will have been *only* a disciplinary tool', or 'political thought will have been *only* a big fiction': we are condemned to managing the economy; there is no longer a political economy. Such statements are attempts at universalising and rationalising the historical failure of a generation of thinkers, militants, and public figures. (TC, 119–20)[8]

A certain amount of this extraordinary desublimatory passage has been lost in translation, though not the term 'abortion' (*avortement*), which is thrown in with quite astonishing political insouciance as a negative marker of non-filiation, as though feminism had never happened. This reactionary natalist gesture is compounded by Stiegler's at best maladroit characterisation of Foucault as 'this great inheritor of Canguilhem's thought'; that is, an 'inheritor' without 'children', or indeed children, of his own, an 'inheritor' whose queer progeny – theoretical, political and personal alike – are very conspicuously absent from *Taking Care* (TC, 122). There is more to this than just the de-queering of Michel Foucault, a common enough move in much recent French-language scholarship on his work. For as James Miller's lurid biography correctly sensed and Mathieu Lindon's recent novel has confirmed, Foucault's distinctively queer practice of *taking care*, whether of himself or of his younger entourage, was one which moved with very considerable ease and relatively little anxiety both up and down the pathways of (de)sublimation (Miller 1993; Lindon 2011).[9]

Before moving on from this outburst of desublimatory fury let me also mention a reference to a note which is placed after the word '*initiation* [*déniaisement*]' and which begins: 'The criminal underworld's term for this is *affranchissement*: an *affranchi* is someone who "knows the score", knows that this is really *only* or *rather* that; for example, this bar is a front for other activities, so-and-so is a police spy, or some influential figure is corruptible, etc.' (TC, 219/217). While Stiegler's intention here seems to be to tarnish *soixante-huitard* discourse of lucidity and emancipation with an air of criminal cynicism, even the reader staunchly determined to remain at the sublimated level of consistencies cannot but be reminded at this point of Stiegler's own incarceration. As is well known, it was during a five-year prison term (1978–83) that Stiegler became a philosopher,

> by chance, thanks to Gérard Granel himself, thanks to the books, the paper, to the knowledge of writing and reading taught to me by the Republic, for which I here give thanks as a child cherishing its mother, through this chance issuing from struggles that were conducted for literacy, notably by philosophers [. . .] I was able to enter philosophy properly speaking by accident, therefore, but also *thanks to the laws of the City*: the spirit of the laws of the French Republic meant there was a library in this old prison. (AO, 22)

I would not wish to make light of Stiegler's gratitude to these individuals, ideas and institutions; nor would I wish to suggest that the work of his philosophical maturity can be read reductively in terms of this foundational narrative of his coming to philosophy, although Ferry, law and literacy are also clearly very much in evidence here. Yet neither Stiegler's abstract reference, with its Greek overtones, to 'the laws of the City', nor his gesture to Enlightenment values embodied in the laws of the Republic, can be allowed to stand without the following desublimatory qualification: seven years before Stiegler was incarcerated, in 1971, Michel Foucault co-founded (alongside liberal Catholic social reformer Jean-Marie Domenach and historian of antiquity and judicial torture Pierre Vidal-Naquet) a short-lived coalition of activists

known as le Groupe d'Information sur les Prisons (or GIP).[10]
While some elements within the group were mainly concerned to
provide shelter for activists from the banned leftist organisation
La Gauche Prolétarienne (GP), the GIP as a whole nevertheless
campaigned vociferously and with some success to improve the lot
of prisoners in France, including on matters such as their access to
news media and other printed material. I shall venture that Stiegler
was himself a beneficiary of the GIP's work and, in that sense, of
the desublimated reality of Foucault's care, as well as of that of the
august institutions, principles and individuals he acknowledges.

Stiegler is rather an adept of the desublimatory gesture; an adept
if not quite a master, for there is a sense in which his fury seems
at times to overwhelm him. The stentorian quality of his writing –
the prevalence of italics, in particular – as well as the domineering
urgency of what Turner calls his 'apocalyptic' tone are likewise
rhetorical strategies of desublimation (Turner 2010: 256) which
sit uneasily with his higher-minded commitment to the cultivation of consistencies. The political question that this performative
contradiction raises but leaves unanswered is: Who gets to wield
desublimatory power in the 'battle for intelligence'? Or, in other
words, according to what principles is desublimatory power to be
distributed? Perhaps this will ultimately come down, for Stiegler,
to a matter of 'revolutionary discipline'; perhaps control of desublimatory power is something that must be reserved for a trusted
elite or self-sacrificing vanguard in the 'organological revolution
of the life of the mind' (TC, 87), a select cohort whose cut and
thrust allows those in their care further down the food chain to
cultivate consistencies. Yet it is difficult to see how such a hierarchised, tactical suspension of democracy could ultimately yield the
democratic results for which Stiegler hopes.

There is a crucial but intractable moment in *Taking Care* when
he contemplates the idea that the battle for intelligence might
already have been lost and that we may accordingly do better to
accept the inevitability of a neo-Platonic caste hierarchy, precisely
the sort of social arrangement that Jacques Rancière's political
writings have done so much to elucidate, even if Stiegler's own
conception of *proletarianisation*, adopted from Guy Debord (in a
détournement of which Debord might well have been proud) adds
a new dimension to those analyses that Rancière did not foresee.
Thus, when Stiegler wonders out loud whether 'it would be better
to cultivate a difference between beings who are mature and thus

organologically armed for the battle of intelligence' and others who are 'like beggars, just good enough to be cannon fodder', all he has to offer in response is this: 'I absolutely do not believe in the truth of this second hypothesis (in fact, it seems intolerable to me). I do not believe that it is rational; I believe that it is *pseudo*reasoning that could be claimed only by an immature consciousness' (TC, 86). What is remarkable about this passage is that Stiegler can offer no sublimated theoretical opposition to this neo-Platonic social settlement and must instead resort to desublimatory strong-arm strategies of insistence even as he is ostensibly struggling to fend off the theoretical prospect of desublimation.

In the final analysis it would appear that the only way to resist the unwanted undemocratic conclusion is through the exercise of argumentative strategies which themselves belong to an authoritarian pedagogical mode. At best this is an aporia; at worst an invidious impasse. Behind the hierarchised distribution of desublimatory power which Stiegler's work implies there is a lingering suspicion that politics is too messy a business to be left to the very people who are called upon to exercise it in a democracy: ordinary people. If, as I have argued Stiegler's practice of theory implies, desublimatory power is to be distributed undemocratically, if it is to be exercised only by a suitably trained, anxiously caring and supremely well-intentioned elite while the rest of the world is schooled in the placid cultivation of social bonds and consistencies, then perhaps the egalitarian response is simply to let go of the sublimatory rope altogether and fall into the *Jauche*, the decidedly messy business of democratic politics.

Notes

1. Preciado 2008: 351
2. 'Civility' is almost certainly more politically and historically vexed a matter than Stiegler cares to acknowledge. Neither he nor any of the inordinately less incisive French moralists who have, acutely after 2005 (and chronically, in the aftermath of 1968), been preoccupied with the 'problem' of its lack have paused to consider the implications of the fact that, as Terry Eagleton puts it: ' "Civilised" conduct takes its cue from traditional aristocratism: its index is the fluent, spontaneous, taken-for-granted virtue of the gentleman, rather than the earnest conformity to some external law of the petty bourgeois' (Eagleton 1991: 32).

3. This may well be a reference that activates a longer circuit than first appears to be the case, via Blanchot, although Blanchot's titular translation is 'Le Très-Haut' and Stielger's 'le plus haut' (Hill 1997: 86).

4. At the risk of further clouding the picture with competing and confusingly imbricated terminology, it is perhaps also worth adding that desublimation (or 'negative sublimation') as Stiegler understands it is roughly equivalent in its theoretical structure and toxic effects to what Marcuse and Žižek term 'repressive desublimation', a concept which could less confusingly have been called 'non-*oppressive* sublimation' because the repression concerned is political (i.e. *oppression*) and not psychoanalytic. Indeed precisely in so far as this is desublimation it transgresses or bypasses repression in the psychoanalytic sense in order, both think, to bolster repression in the political sense (Marcuse 1964: Ch. 3; Žižek 1994: 16). To confuse matters still further, Marcuse also famously advocated what he called 'non-repressive sublimation' (Marcuse 1966: 168), which he sharply differentiated from the explosive and authoritarian operation of 'repressive desublimation' and understood as an expansion of the libido, the 'self-sublimation of sexuality' and the 'reactivation of pregenital polymorphous eroticism' in the absence of, or contrary to, repression in *both* its political and psychoanalytic senses (Marcuse 1966: 165, 172).

5. Chris Turner rightly notes that this is a Ferry whose education reforms are apprehended in disquietingly complete abstraction from the explicitly racist and colonialist terms in which they were invariably articulated and understood, in the first instance by Ferry himself (Turner 2010: 256). Stiegler's is a Ferry of policy rather than politics.

6. While space prohibits a fuller discussion of the sublimated content of this charge, it is perhaps worth pointing out that Foucault *never* said that 'school will have been *only* a disciplinary tool' (TC, 120).

7. 'Reproductive futurism' is defined in Edelman 2004: 19–21.

8. Both of the additions in square brackets are the translator Stephen Barker's in the published translation, which for reasons which will be apparent I reproduce here exactly as printed, although for the sake of clarity without the note references I discuss above.

9. Didier Eribon, by contrast, waited until the third edition of his canonical biography, which coincided with the publication of Lindon's novel early in 2011, before reluctantly lending credence to reports of some of Foucault's desublimatory activities.

10. See Artières, Quéro and Zancarini-Fournel 2003.

IV: Politics – The Consumption of Spirit

11

The New Critique of Political Economy

Miguel de Beistegui

Recently, philosophy from the continent has experienced a remark-able return to questions of political economy, after decades of emphasis on ontology, ethics, and aesthetics.[1] Undoubtedly, the devastating economic and financial crises that continue to shake the world, and of which the subprime disaster of 2008 is only the latest and most acute manifestation, play an important part in such a turn. For the most part, the return of political economy has taken the form of *critique*. In that respect, the return of politi-cal economy also signals the return of critique as such – after a long period dominated by phenomenology, hermeneutics and deconstruction. Stiegler's work since *De la misère symbolique* (*Symbolic Misery*, 2004) and up until *For a New Critique of Political Economy* (2009) can be seen in that light. On the one hand, the author doesn't hesitate to identify the new critique of political economy with 'the very task of philosophy' and to situate his own intervention in the context of the various solu-tions to the crisis considered by the political class, which he dis-misses as unable to identify the real nature of the problem (CPE, 19–20/31–3).[2] Specifically, he emphasises the inability of elected governments and economists to draw lessons from the exhaus-tion of the current, hyperindustrial model of production, which he began to question in *Symbolic Misery, 1: The Hyperindustrial Epoch* (2004). The model in question relies primarily on oil, the automobile and an increasingly vast network of motorways, but also, if not primarily, on an unlimited consumerism. The latter is fed by ever more sophisticated and invasive techniques of mar-keting and communication, which now rely almost entirely on computers and the internet. On the other hand, Stiegler raises the question of critique anew, and of the form it must adopt in the face of the very specific configuration of contemporary capitalism. The very title of the work at issue here, namely, *For a New Critique of Political Economy*, suggests a deep connection with a tradition that goes back to Marx's own critical project. Stiegler pays tribute to Marx's *Contribution to a New Critique of Political Economy* (1859), as well as to the communist journal *La Nouvelle Critique*, which played a significant role in his own intellectual development. But Stiegler's recent work also suggests, albeit subtly, a departure from that tradition by calling not only for a *new* critique, but also a new *sense* of critique. In other words, it raises the question of the form that, today, critique needs to take. Stiegler's contribution, in this respect, consists in a rescuing or affirmation of critique *after* deconstruction, that is, after the system of oppositions and hierarchies constitutive of metaphysics has been exposed and shown to turn against itself. It calls, then, for something like a post-deconstructive critique, or an emphasis on the critical dimension *of* deconstruction. The two aspects of Stiegler's approach are naturally linked, in so far as the crisis of critique stems from the specificity of the current economic and social crisis.

In any case, it is in the context of this question of critique, and of a post-deconstructive critique, that Stiegler's project signals a shift away from Marx and connects with another problematic, inherited from Freud, and another tradition, represented by Reich, Marcuse and, to an extent, Deleuze and Guattari.[3] For the critique of political economy, Stiegler argues, needs to take the form of a critique of *libidinal* economy: the central problem of our hyperindustrial, hyperconsumerist and now increasingly *toxic* form of capitalism is fundamentally a problem of *desire*. From the initial 'Warning' that opens *CPE*, Stiegler posits that the goal is to extricate ourselves from the economic and political model of *consumption*, and construct in its place a new model of social and political *investment*, that is, an investment of the '*common desire*', which he immediately equates with a new *philia*, without which, Stiegler writes elsewhere, 'there is no city-state, no democracy, no industrial economy' (DD1, 15/34). And in the Introduction, the problem of consumption is clearly identified with the economy of desire itself (CPE, 12–13/22).[4] It is by recognising *desire* as the nature of the problem as well as the solution, as the poison as well as the remedy, that the current crisis can be overcome, and critique be given its true and full meaning. Critique itself is to be understood in two different ways, then. On its destructive (and deconstructive) side, the critique of desire takes the form of a critique of the manner in which, through the *system* of consumption, facilitated by a technology of what, following Sylvain Auroux (Auroux 1994), Stiegler calls 'grammatisation', contemporary capitalism has led to the creation of a new, universal proletariat, that is, of a universal class whose only purpose is to *serve* the economic machine. On its more constructive side, the critique of desire aims to revive, or, better still, *invent* a different use of desire, a distinctly *political* desire, that is, a 'desire for the future' (DD1, 24/45; DD3, 82), which Stiegler, following a clue from Derrida, also characterises as a 'desire for singularity', one that is both *incalculable* and *of* the incalculable (DD1, 140–1/186–7).

In what follows, I will try to define and ultimately question the sense of critique that Stiegler is putting forward, whilst emphasising the significance and indeed urgency of a critique of political economy.

The libidinal economy of desire

According to Stiegler, the question with which the *new* critique of political economy is concerned is not, or at least is no longer, that of the working class, but that of the proletarianisation (*prolétarisation*) of labour *as a whole*. It is what we could call the becoming-proletarian of capitalism itself, in which, to use Simondon's terminology, the worker is no longer *individuated* in the system he generates, but precisely *disindividuated*, that is, robbed of his power of individuation. The process of individuation has itself shifted, and is now entirely on the side of the technical object. The 'individual' is thus no longer the worker, but the technical object, or the reproductive machine, to which the worker is entirely *subjected*. The second industrial revolution transformed an economy based on the steam engine, steel and railway networks to one based on oil, the automobile and vast road networks. It eventually gave way to the current 'third' and truly *cultural* revolution, based on computation and the information superhighway, and transformed itself into a *service* industry oriented towards hyperconsumption. The 'consumer', now accessible and mobilised twenty-four hours a day, and whose memory, habits and preferences are entirely retained and stored forever in the digital traces he leaves at every stage of the consumption process, whose identity is entirely externalised in a technical process of *grammatisation*, is the new proletarian. By 'grammatisation', Stiegler means the technical history of memory, beginning with writing itself, which, as Plato already made clear in the *Phaedrus*, is both an aid and a threat to memory. This technical process has now exceeded memory and language and extended to knowledge as a whole (through electronics and digitalisation), bodies, affects and ways of life, notably through the ever-increasing number of services, the creation of 'profiles' and social networks. It has become the process through which the flows that control symbolic or existential acts are rendered discrete, formalised and reproduced. But it also generates a standardisation and formalisation that subjects everything it formalises to 'calculability', thus extending the process of rationalisation which Weber first described, and progressively erasing the incalculable and unforeseeable, or the very possibility of the event. As such, it amounts to a loss of autonomy and recognition, but also of knowledge, both in the sense of the *savoir faire* and the *savoir vivre* implied in the process of technological production. Finally, it disallows the 'psychic' as well as 'collective' individuation in which work *can* consist. At the same time, however, it generates a certain form of socialisation, one in which the 'individual' in the liberal sense is encouraged to pursue desires that have been defined, calibrated and calculated in advance by the market, but in which singularities and the incalculable are repressed. For the process in question coincides with the *systematisation* of consumption and the *fetishisation* of the market through various techniques, such as marketing, advertising and the media, and facilitated, as well as *controlled*, by the internet. This is the reason why Stiegler claims that 'the question of memory and its technicisation' is what is fundamentally at stake in the new critique of political economy (CPE, 36/52).

This decisive evolution also marks the shift from the physical and even mental energy of the worker, now replaced by machines and computers, to the libidinal energy of the consumer – a form of energy that is free and infinite, but also, and for that very reason, ultimately destructive. Desire is the energy on which hyperindustrial capitalism runs, yet it is an energy that turns against itself and sows the seeds of its own destruction: by constantly expanding and accelerating its own growth, notably through an increasingly powerful network of computers, capitalism introduces ever greater and more frequent crises, which risk bringing down the system as a whole. Yet this is a risk that capitalism is ready to run – and in fact has no choice *but* to run – the alternative being a crisis, or indeed the death, of desire itself. 'Desublimation' is the process that turns the libidinal, vital energy into self-destructive drives, and leads to the destruction of capitalism itself, the initial 'spirit' of which was once one of sublimation. It signals the 'tendency of the rate of libidinal energy to fall [*baisse tendancielle de l'énergie libidinale*]' (DD3, 76), or the way in which flows of desire are caught and captured through techno-cultural media:

> The cultural industry and marketing thus target the development of the desire to consume, but, with consumption amounting only to the experience [*épreuve*] of banality, it frustrates and disappoints its desire, *kills* its desire, because it reinforces the death-drive: instead of sustaining desire, industries and marketing cause and exploit the compulsion to repeat. They thereby go against the life-drive. And because desire is essential to consumption, this process is self-destructive. (DD1, 127/169)

This process of 'desublimation', a term which Stiegler borrows from Marcuse's *Eros and Civilisation* (Marcuse 1964: Ch. 3), is nothing other than the 'spiritual misery' (*misère de l'esprit*), or the 'symbolic misery' (*misère symbolique*), which Stiegler also refers to as the 'reign of stupidity' (*règne de la bêtise*, EC, 108–9 [§25]). But Stiegler disagrees with Marcuse on how best to understand this process. Marcuse wants to rescue a golden age of desire, beyond the unfolding of technology, with its struggle for 'domination' and its 'performance principle' – an age that could be reinstated through a revolutionary process, in which the pleasure principle would dominate over the reality principle, and 'instincts' would finally be 'liberated.' Stiegler, on the other hand, emphasises the originary *technicity* and *historicity* of desire, that is, the libido's originary *détournement* or inevitable *perversion* (DD3, 12–13). In other words, on Stiegler's reading, the structural plane is *also* and *from the start* a historical one. The specificity of today's capitalism, in that respect, is to '*divert infantile libido*, normally *invested in parenting, towards the objects of consumption*' (DD3, 76 [emphasis added]). In this late, hyperindustrial age of capitalism, the libido is channelled through, and determined by, consumption, that is, by objects of desire that can be *only* consumed, and kept alive by sophisticated techniques of marketing and advertising, the ultimate aim of which is the total *addiction* and thus self-destruction of the consumer. In that respect, capitalism creates the conditions of its own destruction: as it expands geographically and takes hold of every sphere and aspect of life, it consumes the consumer.

But Stiegler's point is that this capacity of *détournement* is constitutive of the libido and of every psychic, collective or technical individuation: far from being 'natural' or 'spontaneous', its objects are defined by a specific process of collective individuation. This, in turn, means that those perversions are historical phenomena. In other words, the elements of the libido are actually *supplements*: desire is technical or technicised from the start, and it is within a given technical individual that psychic individuation takes place: 'It is not simply civilisation that is the diversion [*détournement*] of libido: libido is its "own" diversion [*détournement*], which is to say its own impropriety: its elementary supplementarity' (DD3, 90). Technics, historicity, in short, grammatisation, far from being the source of the fall of desire, is the process constitutive of desire itself as a '*défaut d'origine*'. As a *pharmakon*, or an originary supplement, it reveals or creates, retrospectively as it were, the missing object of desire, and thus the condition of our own addiction. It is technics, and not the law and its transgression, which creates desire. It is technics as the originary supplement that creates the origin at the same time as it overcomes it, that affirms and negates it at the same time. The *pharmakon*, in so far as it is at once the remedy and the poison, is the 'support of all forms of addiction' or dependency, be that of the fetish or of memory, which gets used to writing as *hypomnesis* (DD3, 115). But the greatest addiction of all, which makes all other addictions possible, is desire: all *pharmaka*, Stiegler claims, are *détournements* of libidinal energy, which means that desire is the energy of all *pharmaka*. Desire is *the pharmakon*. This means that every political economy is a libidinal economy, and that desire is the most toxic as well as the most precious asset. Political economy, then, is the science of libidinal investment.

Once the originary technicity of libidinal energy has been established, the question becomes that of the technical object within which, today, desire is individuated, the specific technical or *retentive* assemblage within which it unfolds. The problem, then, does not lie solely in the historicity of the libido, but in the fact that, in this hyperindustrial or hyperconsumerist age, it inverts itself, and from the vital force that it was becomes a force that leads to the decomposition of the tendencies that constitute it, that is, to their destruction as tendencies constitutive of an individual and collective form of desire. The problem is that it amounts to a '*baisse tendancielle*' of its energy, which needs to be reversed.

Desire and leisure

Now that we have seen how, under the latest hyperindustrial or hyperconsumerist phase of capitalism, the libido's own *détournement* leads to its own destruction or intoxication and amounts to a process of de-composition, rather than composition, the question is one of knowing how this tendency can be reversed, how desublimation can give way to sublimation – how, in other words, the poison can become remedy. Such is the logic and demand of the *pharmakon*. The *critique* of libidinal economy must give way to a more constructive project, which Stiegler characterises as a libidinal *ecology*. It is in the context of such a project that the key distinction regarding desire becomes that of *otium* and *negotium*, rather than the Freudian (or Marcusian) distinction between pleasure and reality. This means that the question of desire exceeds the twofold horizon of the pleasure and the reality principle, but also, and more significantly still, the horizon of the death drive, or the destructive *pulsion* that is the very energy – a toxic one indeed – of hyperindustrial capitalism.[5] But by introducing this distinction, and by isolating a space and time of desire outside the *negotium* of the market, outside, that is, the space and time that are calculable and negotiable, Stiegler does not want to leave the market intact. Rather, he wants the 'spirit' of capitalism itself to evolve and transform itself as a result of that openness to a sphere of desire beyond consumption, and to become a process of sublimation again. He wants the very technology that is currently in the service of the processes of disindividuation to release its individuating potential.

The first thing that needs to be said about the distinction between *negotium* and *otium* is that it does not correspond to the distinction between 'work' and 'leisure'. The reason for that is that, under the current configuration of capitalism, 'leisure' is itself increasingly, if not already entirely, subjected to the demands of the cultural industry: the leisure industry does not aim to free up individual time, but to control it in order to 'hypermassify' it. It actually develops tools for a new voluntary servitude. The distinction between *otium* and *negotium* thus requires, and in fact coincides with, a transformation of political economy itself, in which production and distribution, labour and technology are subjected to a political decision oriented towards a different form of sublimation, conceived as the care of the self and others, the development of new technologies of the self, new processes of grammatisation, exercises and practices that favour an openness to the incalculable and the singular, rather than the reproduction of the same, and the systematic erasure of singularities. This new culture 'is an economy that is as much libidinal as it is commercial, calling for new mechanisms of mutualisation, a new form of government strength, and new objects of social property' (CPE, 51/73–4). It does not signal a return to ancient practices of *otium*, such as those described by Hadot or Foucault in his late work, which can nonetheless help us intimate a different experience of work, involving the creation of one's individuality, and a different experience of time, involving its own incalculability. That such a transformation can involve the convergence of technics and singularity goes almost without saying, given the pharmacological essence of the former: a new industrial society needs to be thought on the basis of a different industrial model, one that involves the socialisation of the technologies generated by grammatisation. What matters, ultimately, is that all such technological supplements amount to a new way of 'taking care' of the fundamental 'value', which is that of 'spirit' or 'desire'. A new form of sublimation can emerge through 'associated technical milieu' that give rise to a new collective, a new 'we' or 'people', a new *philia* (CPE, 50/72). Those technologies of spirit will be developed in place of the technologies of drives, which separate individuals from one another, and they will be developed with a view to an economy of contribution, rather than growth (PFM, 69).[6]

Conclusion

I would like to end this article with a few critical questions. They are all concerned with Stiegler's commitment to Freud's theory of 'sexual desire' or 'libido', and with what I see as a problematic connection with history grammatisation. If, as Stiegler claims, the grammatisation of the libido has always already taken place, whether as sublimation or desublimation; if sexual energy is from the start historicised or technicised, that is, inscribed within technical or technological *dispositifs* or *grammes* that are both individual and collective; if, in short, *eros* has from the start given way to *civilisation*, how can sexuality itself ever be recognised as something other than an *effect* of that historicisation? How, and where, can it be recognised *as such*, in a state of normality? How is it possible to speak, as Stiegler does, of the way in which the libido is 'normally' invested, of its '*détournement*', and of the 'lost' spirit of capitalism? How is it even possible to speak of *sublimation*, if not within quotation marks and with infinite care, and with a degree of scepticism regarding Freud's theory of infantile sexuality, and notions such as 'satisfaction', attachment', and even 'libidinal energy?' Should we not ask whether the discourse of the libido can avoid being engulfed in its own historicity, and call for a *genealogy*, rather than a deconstruction of this discourse? Is the discourse of sexuality itself not a specific object generated within a specific age or epoch – the very age that Foucault analyses in the first volume of *History of Sexuality*? How, when and under what circumstances did the discourse of desire come to be defined in terms of 'sexuality' and 'energy'? The question would become one of knowing *how* and *why* those objects were constituted in the first place. And in that respect, one might wonder whether the very appeal to desire in the terms originally defined by Freud isn't itself determined, if not overdetermined, by a certain energetics and a certain political economy – that of liberalism, and utilitarianism in particular – which Stiegler is precisely trying to overcome. One might want to ask about the *historical* links between psychoanalysis and political economy, and about their profound affinity. If such questions are legitimate, it becomes a matter of shifting the sense of critique away from deconstruction, away, that is, from the pharmalogical or grammatological paradigm, but also away from sublimation as the decisive process of 'civilisation', and towards something like genealogy in the Foucaldian sense. What is in question here is the *sense* of critique, the way in which that word operates in Stiegler's discourse and connects with what I see as a massive *assumption* regarding sublimation as a primary process, and clearly expressed in the following statement:

> All civilisations are ways of capturing what Freud called libidinal energy, diverting it away [*la détourner*] from sexual objects so that it fixes itself onto social objects through which it becomes elevated [*par lesquels elle s'élève*]: a civilisation is a process of sublimation through which the power of the drives is transformed into social energy – above all, to sublimate means to transform. (EHP, 19–20)

One needs to believe, then, in the existence of 'sexual objects' and 'sexual energy' in order to account for the birth of cultural processes, from art and religion to systems of production and consumption. But one can wonder whether those processes, whilst being an expression of desire, all partake of what Freud calls the libido, which is essentially oriented towards satisfaction, or pleasure, and with respect to which the reality principle operates as a force of *différance*. One can wonder whether the desire of writing, to use Blanchot's expression, or the desire of sacrifice, to use Bataille's, are fundamentally driven by a *libidinal* energy, or whether they signal a sense of experience and a form of life 'beyond the pleasure principle', beyond, that is, a primary experience of lack that generates the *Sexualtriebe* themselves. After all, did not Freud himself recognise the existence of this 'beyond' in the seminal article from 1920 in which he introduces the theory of the death drive (*Todestrieb*) (Freud 1955b: 38–41)? Stiegler also recognises it, when he speaks of the destructive drives inherent in hyperindustrial capitalism, but without, I believe, drawing the necessary consequences. Does this horizon, which Freud defines as exceeding the sexual economy of the ego – and thus, at least if we are to follow Stiegler's analysis, economics *as such* – also partake of the logic of the *pharmakon*? Does it function as an originary supplement to the sexual economy of the ego? If so, one would need to recognise an aneconomical, destructive dimension of life itself, and thus a form of desire that would be a threat to sublimation itself, and ask about its role within 'civilisation'. If not, one would need to recognise a play of desire outside pharmacology, and thus the need to extend critique beyond deconstruction. But more significantly still, the question that, beyond Freud himself, and beyond the death drive, would need to be broached is that of the possible overcoming of the capitalism of consumption and hyperindustrialisation through an economy of desire that would not be *libidinal*, and a politics that would not amount to a process of sublimation. For after all, does it go without saying that *philia*, assuming it designates the social link as such, partakes of the libidinal economy, as Freud claims, or could we envisage it as a form of desire that is neither born of a lack, and a missing object, nor rooted in *eros*?

Is it not possible to think friendship, and even love, but also the desire to write, create or think, not as the *transformation* of a *sexual* energy, but as the expression of a *different* energy, as the manifestation of a degree of power, and of a power or force that is not of the ego or the subject, but vaster, more impersonal and *pre*-individual?

Notes

1. Giorgio Agamben, *The Kingdom and the Glory*, trans. Lorenzo Chiesa and Matteo Mandarini (Stanford: Stanford University Press, 2011); *Il Regno e la Gloria. Homo Sacer, II, 2* (Milano: Neri Pozza, 2007); Christian Laval, *L'Homme économique: Essai sur les racines du néolibéralisme* (Paris: Gallimard, 2007); Pierre Dardot and Christian Laval, *La Nouvelle raison du monde. Essai sur la société néolibérale* (Paris: La Découverte, 2009); Bruno Latour and Vincent Antonin Lépinay, *L'Économie, science des intérêts passionnés. Introduction à l'anthropologie économique de Gabriel Tarde* (Paris: La Découverte, 2008); Jon Elster, *Le Désintéressement. Traité critique de l'économie politique* (Paris: Seuil, 2009); Frédéric Lordon, *Capitalisme, désir et servitude: Marx et Spinoza* (Paris: La Fabrique, 2010).

2. Stiegler's claim is that, with a few exceptions (that of André Gorz in particular), philosophy today does not have anything to say about political economy. Whilst I think this claim may have been true, the very increase of publications in that domain is undeniable (see footnote 1). See André Gorz, *Métamorphoses du travail. Critique de la raison économique* (Paris: Gallimard, 2004).

3. This, however, does not mean that Stiegler sees eye to eye with those thinkers. In *Mécréance et Discrédit, 3. L'Esprit perdu du capitalisme* (Paris: Galilée, 2006), for example, the second part of which consists of a reading of Marcuse's *Eros and Civilization*, Stiegler claims that 'the elaboration of a new critique presupposes a critique of Marcuse' (DD3, 72).

4. Stiegler also argues that it's a question of 'engaging in a critique of libidinal economy' (CPE, 40/57).

5. The consequences of such a move are considerable, and perhaps not fully grasped by Stiegler himself. I try and address some of them in my conclusion.

6. Echoes of such an approach, from a Keynesian perspective, can be found in Robert and Edward Skidelsky, *How Much is Enough? The Love of Money, and the Case for the Good Life* (London: Allen Lane, 2012).

12

Stiegler and Foucault: The Politics of Care and Self-Writing

Sophie Fuggle

Introduction

> What, do you imagine that I would take so much trouble and so much pleasure in writing, do you think that I would keep so persistently to my task, if I were not preparing – with a rather shaky hand – a labyrinth into which I can venture, in which I can move my discourse, opening up underground passages, forcing it to go far from itself, finding overhangs that reduce and deform its itinerary, in which I can lose myself and appear at last to eyes that I will never have to meet again? I am no doubt not the only one who writes in order to have no face. Do not ask who I am and do not ask me to remain the same: leave it to our bureaucrats and our police to see that our papers are in order. At least spare us their morality when we write. (Foucault 1972: 17)

This response to an imaginary interlocutor, famously made by Foucault at the end of the introduction to *The Archaeology of Knowledge*, identifies two types of writing that would, in many ways, concern him throughout his career. As a counterpoint to the 'official' writing and documentation of bureaucrats and police officers associated with the production of discourses of truth, Foucault defines his own practice of writing as an exercise in getting lost and, as such, the means by which he might shed pre-existing identities, not least the ones recorded and filed by the authorities. During the 1960s, he would frequently associate writing with the 'dissolution' of self. This was somewhat romantically conceived in terms of a life-shattering, terrifying confrontation of the limits of language and existence heavily indebted to both Georges Bataille and Maurice Blanchot. His late work, which saw him (re)turn to Greco-Roman notions of care or ethics carried out upon the self, gave significant attention to *hypomnemata*, a form of 'self-writing' defined as an 'art of living'.

But where Foucault once wrote in order to have no face, Stiegler's preoccupation is with the way in which we read and write and consequently *have no brain*. As the processes of reading and writing together with the information produced and stored by such processes are increasingly transferred to machines, the cognitive functions once required to perform such tasks are displaced, contributing to an impoverishment of the mind and a subservience to technology that Stiegler describes in terms of widespread *cretinisation*, or what he often refers to as 'bestialisation'. Stiegler employs the term *hypomnemata* to refer to any object external to ourselves which enables the extension of human memory (AH). According to Stiegler, *human* memory has always been 'technical', consisting in our externalisation in technical prostheses. However, in providing an extension of memory, the medium and media through which *hypomnemata* are produced are, at the same time, responsible for weakening such memory.

Hypomnemata constitute a form of *pharmakon* (a term taken from Plato and appropriated by Derrida), at once poison and cure (TC, 34/67). Although a tension has always existed between the two, since the development and implementation of the first tools for drawing and marking (AH), an equilibrium has been predominantly maintained between the loss of one form of knowledge and the development of another. However, a marked shift occurs during the late twentieth century, coinciding with the emergence of *mnemotechnologies* in contrast to earlier *mnemotechnics*. Where *mnemotechnics* are tools that enable memory to be stored externally, *mnemotechnologies* not only store but also generate and organise information, resulting in the 'structural loss of memory' and the 'control of knowledge' by these technologies, which include such devices as televisions and computers.

In taking up the notion of *hypomnesis* at key points in his work, Stiegler offers a useful critique and development of the later Foucault, identifying contemporary applications for hypomnesis with specific reference to the media technologies of late capitalism. This critique should furthermore be situated within Stiegler's radical rethinking of the notion of 'care', alongside his account of psycho-power as the 'apotheosis' of Foucault's bio-power (TC, 13/31–2). Nevertheless, there are various aporias and oversights in Stiegler's work which might benefit from further examination in the light of Foucault's texts on power, care and self-writing. Over the course of this chapter, we will consider how Stiegler redefines power in relation to processes of deterritorialisation that simultaneously produce and are produced by neoliberal modes of production and consumption. Here, we will focus on Stiegler's privileging of time above space, the proliferation of mnemotechnologies and the loss of intergenerational care this engenders. Finally, the essay will consider, as a counterpoint to this loss, the potential of *hypomnemata* for opening up new, alternative modes of care, responsibility and collective action.

From San Francisco to Silicon Valley

In the thirty years since Foucault's death, a number of scholars have occupied themselves with redefining power in relation to late twentieth century, neo-liberal society. Gilles Deleuze's oft-cited 'Postscript on Societies of Control' constitutes perhaps the best-known example of such an attempt to redefine power. In identifying a shift from disciplinary to control societies, Deleuze presents us with an account of the increasing prevalence of information technologies in producing, defining and regulating individual subjectivities. Individuals give way to 'dividuals', composed of usernames, passwords and pin codes (Deleuze 1995: 182/243). It is important to note here the influence the text has had on Stiegler, who locates the development of mnemotechnologies en masse at the (decentred) centre of present-day control societies (AH). Yet, where Deleuze's essay has assumed notoriety beyond the scope of its own limited critical analysis, Stiegler provides perhaps the most complex, sustained and original account of post-disciplinary power.

Although Foucault himself had begun to call for a rethinking of power beyond the disciplinary, he nevertheless continued to view disciplinary power as a contemporary form, stating in a 1978 interview that 'we need to detach ourselves in the future from today's disciplinary society' (Foucault 2001b: 533). Stiegler, on the other hand, firmly aligns biopolitical power with nineteenth-century industrial Europe, claiming that power in the twentieth century can no longer be considered as directed at populations occupying given spaces. Here he seems to be echoing Foucault's underdeveloped notion of 'security', set out in *Security, Territory, Population*, but not followed up elsewhere. Security constitutes a mode of power emerging alongside yet in contradistinction to the disciplinary, aimed at growth and circulation rather than control and regulation (Foucault 2009: 44–5). As such it prefigures Stiegler's account of psychopower, defined as a new set of technologies belonging to an era of deterritorialisation in which individuals are constituted not as workers but, rather, as consumers. The lives and actions of individuals are thus no longer constrained by the nations or states in which they live but subjected to the whims of the economic market, composed of deterritorialised flows of capital (TT1, 86; TC 55–6/106).

Of key importance here is the shift from a focus on power in terms of its spatial configuration to its operation on the temporal. This is a move that can already be mapped onto Foucault's work despite his claiming at various points to be rehabilitating the spatial as an object of philosophical enquiry that, since Kant, has been subordinated to questions of time (Foucault 1980: 149–50). The juxtaposition of the torture and execution of Damiens and the prison timetable set out in the opening chapter of *Discipline and Punish* still remains today one of the most powerful prefaces to a critique of social institutions. It also exemplifies the temporal and spatial constraints placed upon the individual body (Foucault 1995). Yet, while the length of the torture to which Damiens is subjected cannot fail to escape our notice, it is the space of the body and its specific location in public space that constitutes the primary focus in Foucault's gruelling account. Conversely, where the prison timetable indicates the various locations occupied by the inmates throughout an average prison day, it is the temporal structure of the day, with no minute left unaccounted for, which is of key importance here. Likewise, when we juxtapose either nineteenth-century institutional power or even the Fordist workplace of post-war US and Europe with contemporary forms of social organisation, what becomes apparent is the shift from a worker occupying a specific space, the production line, desk or counter, to a worker supposedly emancipated from the confines of the workplace only to find that the once-strict parameters between work and play or work and home have now become eroded to the point that one responds to emails from the boss whilst on the bus, on vacation and even in bed.

Thus, it is the question of time rather than space that takes primacy. The disappearance of clear spatial limitations between workplace, home and what Starbucks (following Habermas) refers to as the 'third place' is considered largely in terms of the increased temporal constraints placed on individuals. If there is no longer any threshold to cross between work and leisure, no factory gates to enter, then there are also no limitations to the working day. So, it appears that it no longer matters where our bodies are but, rather, how our time is being spent.

It is this question of time which is taken up by Stiegler. In *Technics and Time, 1*, he maps out a post-Heideggerian project, identifying time as that which both produces and determines being. Our identity *qua* human is described as consisting in an internalisation of our external prostheses. We exist in a state of differance, suspended between the technical prostheses that we are now, and those through which our interiority will be reinscribed in the future (TT1, 138–40). Elsewhere, Stiegler's focus on time involves a rethinking of Foucault's work on power and care of the self, particularly in *Taking Care of Youth and the Generations*. Of fundamental importance for Stiegler is what he refers to as a process of proletarianisation, which is no longer defined by class boundaries but rather constitutes a post-Marxist critique of technology and the machine. Where once we were alienated producers, unable to consume the profits of our labour, we are now alienated consumers, deprived of the possibility of producing what we consume – denied the prospect of creating for ourselves the prostheses in which our interiority consists. Worked by and for the machines that invent us, our servility to contemporary forms of technology extends beyond the limits of the working day and reconfigures what time itself means. Here, Stiegler draws upon the notions of deep and hyper attention defined by Katherine Hayles (TC, 73–8/137–41). Our ability to sustain concentration on one single intellectual activity over a period of time (deep attention) has been replaced by hyper attention, which sees us flit from one activity to another without maintaining concentration on any single activity for more than a few minutes, and often only a few seconds. This shift in attention might enable us to carry out multiple tasks simultaneously but obviously affects our capacity for sustained intellectual reflection (Hayles 2007: 187, cited in TC, 73/137).

If the body, for Foucault (and Nietzsche), constituted a battlefield upon which war is waged between competing forms of knowledge (Foucault 1977), the stakes have changed. Of interest to Stiegler is not the site of the battle – moving his focus from the body *per se* to the cognitive – but, rather, the battle itself. And, it is no longer a question of competing discourses and knowledge but knowledge itself. This is what Stiegler refers to throughout *Taking Care* as the 'battle for intelligence' (see, especially, TC, 17–22/37–47). We find ourselves caught up in a struggle in which the so-called 'knowledge societies' and 'knowledge industries' that define late capitalism produce and celebrate the very apparatuses and technologies responsible for widespread loss of knowledge, perceived in terms of the attention and retention of memory (AH, TC).

However, in analysing the battle as a battle for intelligence, we cannot completely do away with the battlefield, which is still the body. Coupled with the deterritorialised flows of capital that define neoliberalism are complex processes of nation-state building and border control. The disciplinary organisation of space also persists within the walls of prisons, schools and hospitals. Stiegler's account of psychopower needs to be combined with Foucault's anatomo-politics if we are to avoid the reductive and, indeed, potentially dangerous gesture of reducing all social existence to the purely temporal. Nevertheless, recognising that processes of deterritorialisation are intertwined with disciplinary mechanisms is not to deny the effects of such deterritorialisation on social relationships, and in particular, the breakdown of individual responsibility and 'care' towards others.

Your parents are *not* your friends

While both Foucault and Stiegler identify a loss of care at the centre of their work on power operations, they define 'care' in very different terms. Foucault opts for the word 'souci', which can also be translated into English as 'worry' or 'concern'. In this respect, it is a care that is essentially inward-looking and constitutes a form of knowledge or truth. The care that Foucault is preoccupied with in his later work is a care carried out upon the self. While this involves a series of practices and activities both physical and reflective, these are all concerned with individual subjectivity, one's capacity to develop an ethical and, at the same time, aesthetic framework in which to live. Thus, for Foucault, care is a form of self-knowledge to be distinguished from the types of knowledge associated with the confessional practices which emerge in early Christianity and are subsequently transposed into other forms of institutional power – judicial, medical and educational.

Stiegler, on the other hand, uses the term 'soin'. The verb 'soigner' can be more accurately translated as 'to care for' and in contrast to 'souci' implies a more active, less reflective form of care. In *Technics and Time, 1*, 'care' is translated from Heidegger's *Bersorgen* in terms of a projection towards the future which attempts to anticipate or determine future possibility (TT1, 6). In this respect, it appears to bear more in common with 'concern' or 'souci', particularly with respect to Stoic notions of training (*gumnazein*) and preparation for the future, which played a significant role in the care or ethics of the self (Foucault 2005: 425) However, in a short paper entitled 'Take Care', Stiegler pursues the notion of 'care' from a different angle – one that seems to be more relevant to his discussion of responsibility towards others than to individual self-awareness and mastery. Stiegler articulates the notion of care and, more specifically, what it means to 'take' care as a political question:

> Here, then, is a task of political economy which echoes similar questions found throughout the rest of society. A lack of political thought about these questions means we are seeing the development of situations of conflict, where those who wish to enter into the service of this cult which is culture, including agriculture, those who would like to take care, set about rejecting the techniques which are precisely, however, the therapeutics of such a care. (2008a)

In this respect, his reading of care should be offered up as a critique of Foucault's account, which affirms the moment when the term is emptied of any political dimension.

Delineating the Hellenistic period as a veritable 'golden age' of a *culture of the self*, Foucault (2005: 30) highlights the moment when care becomes detached from its association with political power. Care of the self ceases to be a form of training in self-mastery undertaken by adolescent males destined for positions of political authority. In focusing on this period where care of the self becomes an end in itself, Foucault is not proposing a contemporary replication of the practices and techniques carried out in first and second Greco-Roman society. Rather, he is identifying these as a particular example of how subjectivity might be formed and developed through practices distinct from dominant social discourses and outside the institutions which house them:

> When, in Hellenistic and Roman culture, the care of the self becomes an autonomous, self-finalised art imparting value to the whole of life, is this not a privileged moment for seeing the development and formulation of the question of the truth of the subject? (Foucault 2005: 254)

However, in his celebration of this moment, Foucault often seems to be celebrating an apolitical, narcissistic mode of existence. Despite appearing to be opened up to all, the possibility of practicing a 'care of the self' continues to belong to the elite. Moreover, his neglect of the higher aims and purposes of Stoic techniques and practices has led Pierre Hadot among others to accuse Foucault of applying a twentieth-century form of 'dandyism' to Antiquity (Hadot 1989: 267). Foucault has stated on various occasions that the 'care of the self' he is referring to should not be understood in terms of either a 'moral dandyism' or what he refers to as the 'Californian cult of the self' (Foucault 2005: 12; 2001b: 1222). Yet, if there is no higher end than the beautiful life, lived in and for itself, then surely dandyism or 'Californication' are precisely what Foucault is proposing. The issue regarding these styles of living and the notion of self-transformation proposed by Foucault is that they have already become absorbed by dominant discourses on identity and individuality and reproduced within the advertising and marketing strategies which form part of the mainstream power grid. Stiegler locates this problematic within a wider critique of (post)structuralism's failure to think through the 'libidinal economy' of capitalism. For Stiegler, we cannot simply condemn capitalism but, instead, must conduct an active critique of its ability to recuperate that which resists or sets out to negate it (MS, 108).

Consequently and in contrast to Foucault, Stiegler identifies what is lost when a system or structure of care breaks down in late capitalist society. Care here is synonymous with the notion of 'cultivation' (Stiegler 2008a), a process whereby knowledge and skills are imparted from an older to a younger generation. This process embodying 'care' also involves a necessary element of violence. Stiegler takes his cue from the link between 'cultivation' and agriculture, which enacts violence upon the natural, which is tamed, contained and harvested at the same time as it is tended and nurtured (2008a; see also 2008b: 20–1). It also implies a framework of trust or belief in those assuming a duty of care by its recipients. With the advent of psychotechnologies, a rupture occurs whereby the younger generation not only surpass their elders in their ability to master new technologies but in doing so lose the capacity to engage with the older generation, who are henceforth deprived of their responsibility and duty of care. This abdication is reinforced through parents' attempts to be 'friends' with their children rather than acknowledge the necessary hierarchy of their relationship.[1] The consequence of this is, at the same time, a loss of trust in the processes of knowledge transmission. Stiegler attributes this to the 'all-out' privileging of technological innovation not supported by what he terms the 'fabric of trust, of fidelity, of belief, of socialisation and of individuation' (2008a; see also DD2). A process of flattening occurs, eliminating the distinction between adults and minors, positing them all as 'consumers' (TC, 1–3/11–15). At the same time, Stiegler identifies the shift from the production of the individual from infancy as a 'docile body' and compliant member of the workforce to the creation of a complex consumer whose individual peculiarities are now actively encouraged rather than regulated, as a result of the ever-increasing possibilities for consumption they enable.

Yet, it is not without a degree of irony that Stiegler locates a duty of care in the nineteenth-century educational apparatuses criticised precisely for their lack of care by Foucault, who identifies them with normalising processes that produced knowledge about a subject not in order to nurture and develop individual subjectivity but rather to regulate, cure and control. Worth noting here is the link Stiegler makes between nineteenth-century educational institutions and the development of mass printing techniques, something he also flags up as being glaringly absent from Foucault's work. Yet, in identifying this important omission and laying the groundwork for further investigation into the relationship between what he calls *grammatisation* and disciplinary institutions (legal as well as pedagogical), Stiegler openly endorses the work of Jules Ferry and Third Republic education reforms in general (TC, 63/118). As Chris Turner points out, this affirmation of newly democratic forms of education fails to acknowledge the engrained racism and elitism of Ferry's reforms. Worse perhaps, is the reactionary and perhaps unwitting nostalgia Stiegler demonstrates for an age before audio-visual technologies invaded the classroom, which Turner aptly mocks for its apocalyptic hyperbole (Turner 2010: 256).

Read thus, Stiegler might seem to be advocating the traditional family unit as a key organising principle for contemporary society. As Foucault points out in *Psychiatric Power*, the family has always functioned as a microcosm of the dominant mode of power, transforming itself from the embodiment of sovereign power in the form of *patria potestas* into the most effective, extensive manifestation of disciplinary power, a central point or nucleus at the intersection of all other forms of institutional power (Foucault 2006: 81). Consequently, it is important to consider how the notion of 'care', at odds with disciplinary, knowledge-producing forms of power, can be located within the space of the family and furthermore how a duty of care might be assumed by an older generation in order to engender alternative social and political configurations to those in play in late capitalism. Here, Stiegler's call to 'take care' cannot, and must not, be reduced to nostalgia for an earlier, seemingly less noxious form of capitalism. A more nuanced reading is required which rethinks the family in terms of a series of complex relationships which might provide the basis for thinking through the organisation, responsibility, authority and, most of all, care required for larger collectives.

Moreover, in thinking through what form such care (*soin*) might take, we should avoid simply calling for a reinstatement of pedagogical practices that deny rather than incorporate new forms of technology. Stiegler is not advocating a blanket dismissal of such technologies as toxic. Such a dismissal would fail to appreciate their role in forging and developing new forms of individual and collective subjectivity, subjectivities which do not preclude but rather enable alternatives to dominant forms of power. It is also worth reiterating Foucault's identification of the important relationship between care of the self and knowledge of the self. This relationship must necessarily bring itself to bear on the processes whereby knowledge and skills are passed from one generation to the next. Again, to view this in terms of nineteenth-century educational practices is to miss the point – what we end up with here is an understanding of knowledge and its transmission which echoes the 'little vessels' and the 'hard facts' of Charles Dickens's *Hard Times*.

The first question with which philosophy should be concerned, Stiegler tells us, is not the law, power or even poetry but, rather, teaching (TC, 107/195–6). Moreover, in describing the process whereby he became a philosopher 'by accident', during his incarceration at the end of the 1970s, Stiegler himself shows how technics can serve as an antidote to power. He identifies the fundamental role of *hypomnesis*, the hypomneses of 'books read and words written' (AO, 18), in reconstituting every day the exterior world needed to counter the interiority of prison existence. It was this process which not only assured his survival but at the same time became his 'philosophy' – 'a pure fabric of *hypomneses*, of which I daily deposited traces on paper, like a snail sliming along a wall' (AO, 19). The final section will explore the notion of *hypomnesis* as a political as well as philosophical project.

The (impossible) politics of self-writing

Stiegler's claim that mnemotechnologies have produced a flattening whereby we are all subject to the same processes of 'proletarianisation' forms part of a larger attempt to rethink class, subjectivity and technology within neoliberal society. However, to fail to distinguish adequately between a corporate CEO being worked by the machine and a migrant worker surviving on casual, precarious labour would be reductive and potentially damaging to Stiegler's political project. In 'Anamnesis and Hypomnesis' he states that: 'the question of hypomnesis is a political question, and the stakes of a combat: a combat for a politics of memory, and more precisely, for the constitution of sustainable hypomnesic milieu' (AH). A reverse and equally problematic move occurs with Foucault's reading of care of the self as something which becomes, during the Hellenistic period, open to all. A latent elitism is operating at the heart of both narratives. When thinking through the process of proletarianisation identified by Stiegler, we should bear in mind that so-called 'ubiquitous media' are in fact fully accessible only to a privileged (Western) portion of his 'deterritorialised' globe. An important tension exists here between the way in which the majority of consumers of technology are bound and defined by these technologies and the very specific, complex consequences this process has on individuals belonging to different social classes and inhabiting different cultural spaces. Foucault acknowledges the inherent elitism of a mode of living which is only open to those with the time and money, yet by actively refusing to draw comparisons with equivalent practices belonging to late capitalism, he fails adequately to criticise the notion of care as a complex mode of 'truth' formation. Again, I want to suggest that reading Foucault and Stiegler together rather than at cross purposes holds the key to a more fruitful discussion of power and care within neo-liberal society. To achieve this I will return to the theme of writing which opened the chapter in order to consider the role of self-writing as engendering rather than simply precluding collective forms of action and debate.

Foucault locates the practice of self-writing within a more general 'art of living' found in first and second century Greco-Roman society. Moreover, he defines it as a mode of *askesis*, the training of the self by oneself. Although writing is a relatively late addition to the 'art of living', Foucault indicates the importance with which it was regarded even by those privileging the oral tradition. His focus is less on the text being produced and more on the process involved, the (circular) 'act' rather than the (linear) 'product' of writing, which accordingly becomes an exercise in meditation and training. Foucault highlights its 'ethopoietic' function as an 'agent of the transformation of truth into *ethos*' (Foucault 1997: 209).

The physical inscription of power on the body and in the architecture of disciplinary institutions is thus central to Foucault's discussion. Stiegler nonetheless identifies Foucault's failure to think through the physical and material act of writing as a key oversight. How, Stiegler asks, can Foucault be at once so interested in the role of the law in defining and producing power relations yet fail to acknowledge the processes of grammatisation that make the writing of 'law' possible in the first place? (TC, 124–5/224) Where Foucault would call for a renewed focus on the body and the spaces it inhabits alongside Stiegler's consideration of the temporal effects of psychotechnologies, elsewhere it is Foucault's neglect of the technical apparatus – the tools, systems and surfaces Stiegler highlights as making writing possible – in favour of the 'act' of writing that needs to be redressed.

In positing personal notebooks and correspondence as two primary instances of 'self-writing' belonging to an 'art of living', Foucault provides a very precise and context-specific definition of *hypomnemata*. Stiegler, by contrast, offers a wider application of the term, situating it within a genealogy of technics and knowledge-production which charts the loss of its ethopoietic function. However, it is worth thinking through Foucault's account of *hypomnemata* as a kind of self-writing that would enable the inscription of alternative modes of subjectivity, identity and interpersonal relationships at a remove from those produced via the institutional apparatuses associated with disciplinary power and the mass media technologies Stiegler identifies with psychopower. Considering these specific practices also enables us to explore further what Stiegler means when he calls for an 'ecology of associated hypomnesic milieu' (AH).

Hypomnemata, conceived in the narrow context proposed by Foucault, consist of notebooks and other personal texts in which key quotations and ideas are jotted down for future reference (Foucault 1997: 209–10). As indicated above, however, their function as memory support is decisive, since for Foucault and Stiegler alike neither writer nor material is left unaffected by the process of writing. Foucault describes *hypomnemata* as attesting to a fundamental tension between the established, authoritative discourse which constitutes the primary material that is read, noted down and reappropriated by an individual, and the specific relationship this process has to that individual's care of the self. Various intertwined processes embody this tension. The act of writing is encouraged alongside that of reading, because it both acts as an organising principle for the ideas encountered and prevents the confusion and distraction that occurs through excessive reading without reflection. Furthermore, the production of *hypomnemata* affirms the importance of wide, disparate reading rather than focus on a single author, text or idea, since what is at stake in this type of writing is the ability to bring diverse elements together into an overarching narrative. As a result of this organising principle, a two-way process is in play whereby the individual puts his particular stamp or identity onto the texts being read at the same time as the texts produce his identity.

This dual process, a form of what Foucault refers to as subjectivation, in contrast to his earlier accounts of subjection (*assujettissement*), is defined by Stiegler in terms of Simondon's notion of individuation. For Stiegler, following Simondon, the relationship between the individual and his or her technical prostheses is 'transductive' and 'ontogenetic', meaning that the identity of an individual emerges from his or her relationship with technics; it does not precede this relationship (TT1, 17–18; see also Simondon 2007: 10). *Hypomnemata* as conceived by Foucault similarly implicate the individual in the production and development of his own identity, which is situated within yet differentiated from the dominant discourses of a specific socio-historical moment. For Stiegler, as for Simondon, however, the process of individuation is diminished with the advent of analogue mass media in an age of 'industrial democracies', where the machine is still able to transform the individual, but the individual has little role to play in transforming the machine (DD1, 57–8). Most notable in this respect is television, whose programming schedules structure and organise the dissemination of information within society, turning citizens from voters into audiences (TC, 52–3/101–2). Television is just one example of how mass media short-circuits (trans)indi- viduation, by making for us the selection of what material we read and internalise. We see the takeover of the excessive, unfiltered reading (watching) that Foucault, borrowing from Seneca, terms 'stultitia' (Foucault 1997: 211).

With the shift from analogue to digital forms of media, two distinct possibilities occur. In the evolution from mass to ubiqui- tous media, the temporal constraints are lifted and individuals are invited actively to construct their own interaction with the multi- ple screens now framing their existence and experience. First, this process constitutes a form of hyper-individualism associated with Deleuze's control societies, in contradistinction to the normalising processes of earlier disciplinary societies. Self-writing, the consti- tution of the self via the act of producing commentary on the self, becomes associated with a loss of self. Where once it belonged to an 'art of living' in which an individual developed a critical space separate from the prescriptions of dominant legal and political discourses, in late capitalism, it attests to what Nealon has referred to as an intensification of power relations, whereby the individual *qua* consumer is actively encouraged to differentiate him or herself via perpetual self-commentary (Nealon 2008). Individuals are given the semblance of greater autonomy, in the form of proliferating media, which offer unprecedented possibilities for self-commentary, in particular, mnemotechnologies that map our locations, send automated responses and invite us to self-define in terms of what we 'like'. These technologies thus undertake com- mentary on our behalf, enacting a form of self-policing that is the apotheosis of Bentham's panopticon.

The second possibility presented by Stiegler is that such media, given the right circumstances, can open up possibilities for renewed forms of (trans)individuation. Where Foucault's defini- tion of *hypomnemata* is concerned with individual care as an end in itself, his account of correspondence is perhaps more useful as a framework for thinking about how the internet and specific forms of social media provide genuine opportunities for collective organisation and action. Foucault points out that correspondence as a mode of self-writing should not be considered merely as an extension of *hypomnemata* (understood here in its narrowest sense). It is more than simply a form of training since it also constitutes the means whereby one manifests oneself 'to oneself and to others' (Foucault 1997: 216). Unlike other practices carried out upon the self, correspondence represents a particular exercise through which an individual subjects himself to the gaze of the other and thereby articulates his position in terms of his relationship to them.

The ability to offer running commentary on the banal activities of our family, friends and colleagues demonstrates the ease with which we become subsumed in everyday minutiae that circumvent sustained reflection and action. Likewise, the ease with which we can provide links to articles and other electronic texts epitomises memory support turned memory loss: in place of the individuating process of taking notes, we merely cut-and-paste weblinks without a moment's thought. Yet, at the same time, critical spaces are opened up outside traditional institutional domains – the school, university and mainstream press – in the form of blogs and online forums. While there remains the issue of 'quality control' and how widespread access really is, the internet, and this is the point Stiegler is making towards the end of 'Anamnesis and Hypomnesis', has the potential for a renewed 'care' of the self. This is a 'care' perceived in terms of Foucault's definitions of self-writing, hypomnemata and correspondence, a 'care' that whilst operating within dominant neoliberal frameworks, goes some way in removing the elitism that hampered Foucault's work on the Greeks.

Conclusion

Stiegler's critical engagement with Foucault is largely predicated on the process of reading Foucault against himself, so as to suggest that his very specific account of *hypomnemata* as a practice developed within a certain socio-historical framework might nevertheless be mapped onto twenty-first century mnemotechnologies. At the same time, Stiegler's redefinition of 'care' goes some way to articulating what a contemporary 'politics of care' might look like, something which Foucault failed to do in his study of Greco-Roman practices associated with a 'care of the self'.

Nevertheless, despite attempts to suggest how a space of resistance might be carved out of contemporary configurations of power, doubts persist as to the extent of this resistance. These doubts echo those raised by Foucault himself, when qualifying his infamous claim that 'where there is power, there is resistance' with the acknowledgement that resistance is always strictly relative (Foucault 1990a: 95). Even where tools like the internet and the social media it houses are deployed in the service of alternative politics at the level of what De Certeau would call 'tactics' (Certeau 1988: xix), the potential for radical reconfiguration of social structures and the technologies framing them is perhaps more limited than one might hope.

Note

1. This point was made by Stiegler during a lecture given at Goldsmiths in February 2011 in relation to the more general question of the possibilities of 'friendship' in a society saturated with social media.

13

Technology and Politics: A Response to Bernard Stiegler[1]

Richard Beardsworth

Bernard Stiegler gives a radical account of technology that shifts the ground of recent critical theory in important ways. I wish in this article to do two things: to show its importance and its limits. To this end, I begin by giving a brief overview of Stiegler's account, emphasising four major shifts from critical theoretical debate of the 1990s. I then argue for the contemporary importance of political economy for critical thought and suggest that Stiegler's recent 're-writing' of Marx's critique of political economy addresses this importance, but is misplaced given undue emphasis on the technological determinations of present capitalist forms. I suggest that this undue emphasis constitutes technological determinism and leads to too totalising and bleak a reading of capitalism. Stiegler's re-writing of critical political economy is also grounded on Freudian libidinal economy (following Felix Guattari and Gilles Deleuze and Jean-François Lyotard), which the following section analyses. I suggest that this reading, while imaginative and provocative, is also technologically determinist: it runs the risk of a unilateral reading of affect in contemporary capitalist forms and thereby limits cultural play. Due to these two forms of technological determinism, I conclude that Stiegler offers us at one and the same time a ground-breaking philosophy of technology and a frozen notion of political possibility.

The technological thesis

The initial singularity of Bernard Stiegler's work is to be found in his use of the work of Jacques Derrida. It takes the grammatology of archi-writing and develops this grammatology into a general philosophy of technology (see DT). This philosophy simultaneously radicalises Husserlian phenomenology. It places the

Husserlian analysis of temporal ecstasis in a field of spatial materiality that is neither Heideggerian nor Derridean and allows for
Simondon-type theses on individuation processes to enter critical
theory (see TT1, 83–276/191–279). In this interdisciplinary field,
the condition of the *epokhē*, ecstasis, and/or trace becomes the
technical support from which psychic and collective individuation
proceeds. This condition, while structural to human finitude (what
Stiegler calls 'the originary default of the human'), is composed of
nothing but the history of memory-supports that underlie hominisation and civilisation – that is, the human animal's increasingly
complex adoption of its natural and technical environment. The
industrial revolution and the capitalist organisation of technology
lead to the institution of the market economy and specific rules
of valourisation. From the above philosophical and genealogical
perspective on technology and its constitutive role, the industrial
supplement opens up a new epoch of human finitude.[2] The specificity of this supplement only comes to the fore, however, in the
last quarter of the twentieth century with the increasing convergence between objects of the mind and objects of production and
consumption – what is called today 'cognitive', 'informational' or
'hyperindustrial' capitalism.

Now, for Stiegler, each epoch within the history of the supplement has required cultural and political appropriation in order for
humans to 'redouble' the movement of technicity as a movement
of 'spirit'. Monotheism represents, for example, the religious–
political adoption of non-orthographic writing, and the theory
and practice of the social contract presents a political incorporation of the printed, alphabetical word. For Stiegler, the hyperindustrial support requires, but awaits its own political adoption.
It is therefore, and above all, due to present failure to adopt contemporary forms of technology that complex modern societies
are becoming de-politicised. In much contemporary discourse, de-
politicisation ensues from economic and legal over-determinations
of society. For Stiegler, conversely, these processes of depolitici-
sation are at root a problem of radical technical change. In the
history of the supplement ('epiphylogenesis' in Stieglerian terms)
the technical novelty that contemporary humankind must address
is the above convergence between processes of the mind and the
logic of the market (cognitive capitalism). This stage of capitalism
increasingly blocks individuation processes and stymies the possibility of a new form of 'spirit'. Indeed, in supplanting traditional

or familial markers of identification with the new technologies, the industrial supplement has come, during the last thirty years, to foster increasing disbelief, cynicism, resentment and despair.[3] For Stiegler, the political adoption of contemporary forms of capitalism constitutes, therefore, an urgent historical task. The present form of the technical supplement, together with the singularity of convergence that accompanies it, specify in turn the terms of this urgency.

As I argued in several pieces in the mid-1990s, Bernard Stiegler's philosophy of technology shifts the ground of critical theory in important ways (Beardsworth 1996a: 145–57; see also 1995, 1998b). I foreground what I still consider today the four most important. First, in Nietzschean vein, Stiegler sidesteps the totalising Heideggerian gesture around technology by giving a dynamic and rich genealogical history of technical supports, one that allows us to understand precisely how humans appropriate technology for their own cultural ends. This genealogy ultimately places Stiegler's work outside the terms of deconstruction. Stiegler's move remains an important reflection on, and conversation with, the work of Jacques Derrida in this respect. [. . .]

Second, the originality of this critical philosophy of technology lies specifically in its exposition of 'tertiary memory'. Husserlian phenomenology is concerned to show the way in which the temporal ecstases of past, present and future are held together through retention and protention in order for there to be a past, a present, and a future in the first place. Stiegler argues that Husserl leaves out of this account the exteriorisation of memory in technical supports. The point is not empirical, but structural. Although external memory supports are empirical derivatives of human memory, they constitute, from the first, the way in which we conceive the past, present and future. Human animals, we are *a priori* addressed by technical supports (from birth, through education, to information-access, etc.). Stiegler calls this form of memory or retention 'tertiary memory'. It follows empirico-historically 'primary' perception in the present and 'secondary' retention of the present as past; but it precedes both structurally. Whatever the epoch of civilisation, the technical support constitutes a pre-given inscription of memory from which humans temporalise themselves and their world. This theoretical addition to Husserl inscribes modern phenomenology in a materialist history which, at one and the same time, undoes the quasi-formalism of the Derridean 'trace' and, as we shall see, proposes a re-writing of Marxist themes compatible with cognitive capitalism. I consider this innovation both paramount and problematic. It provides a new technological determinism that considers technology radically constitutive of the human condition; at the same time it risks unilaterally reducing this condition to its technical supports.

Third, Stiegler's concern to place critical thought in the present is strong. While Stiegler's formulations are often totalising and apocalyptic, his sharp sense of political purpose remains important. After twenty years of post-structuralist reflection on the so-called 'withdrawal of the political', Stiegler was right, for example, to recall critical theorists from the 1990s onwards to the outstanding questions of politics: 'Who are we?' and 'What do we want?' His reminder was, for me, convincing because it was neither structurally prescriptive nor arbitrary. It followed immanently out of a reflective history of tertiary memory and a consequent analysis of the requirements of politico-cultural adoption. There is thus, for Stiegler, little need to suspend this 'we' in indeterminate and self-conscious scare-quotes – out of fear of a politics of identity. Read from the perspective of 'tertiary memory', contemporary history cannot bring necessary processes of identity-formation through technical supports to a postmodern end.

As Stiegler rightly emphasises – contra deconstruction, in particular, and postmodernism, in general – political appropriation and ownership constitute a condition of material processes of civilisation. Stiegler's philosophy of technology has, therefore, tabled in a singular manner how important, and how problematic, this 'we' is today in a globalised world. The gap between globally organised technical supports and their political adoption may indeed be the political challenge of the future. However impractical this future seems, Stiegler's philosophy of technology has shown, at most, the need for a global politics (although he has only addressed this politics in general terms) and, at least, the precariousness of recent distinctions between modernity and postmodernity. Fourth, and very briefly, Stiegler's philosophy of technology has argued for reason and for institution at a moment in critical theory and postmodern debate when there remains a strong and debilitating suspicion of the dominating tendencies of post-Enlightenment reason and its institutions. His philosophy of technology argues well, I believe, for rationally substantive political invention in this intellectual context. These four traits have distinguished Stiegler's work from recent critical theory and place it at the potential vanguard of contemporary explorations of critical political and cultural analysis. That said, there are important problems in Stiegler's manner of understanding contemporary reality and political adoption. I believe it important to highlight them within the ongoing process of cultural translation of Stiegler's work. I will address them through his reengagements with two thinkers: Karl Marx and Sigmund Freud. For the way in which his philosophy of technology interprets their legacy foregrounds the manner in which his understanding of reality must be in part contested. [. . .]

Stiegler and Marx: the contemporary question of Political Economy

I argue above that Stiegler's conceptualisation of technology is made through a radicalisation of the Husserlian understanding of retention and protention. Stiegler uncovers and focuses upon the specificity of tertiary retentions: the technical support constitutes a pre-phenomenal inscription of memory from which humans temporalise their world. This move reconfigures historical materialism. Marx's distinction – found most succinctly in the *German Ideology* – between modes of production and relations of production mis-recognises, for Stiegler, three aspects of technology when considered as technical support. First, since technology is constitutive of hominisation, it cannot constitute *per se* a means to a human end. The Marxist theory of production remains instrumentalist and humanist. As a result, Marxism may underestimate the difficulty of political adoption (or at least conceive of it in exclusively class terms). Second, the Marxist distinction between the matter of technology (modes of production) and the manner in which it is socially organised (the relations of production: most importantly, property relations) is superseded by the original conjuncture that the technical support provides between the individual and the collective (see particularly DD3, 49–59). Technical supports, as historically mediated forms of memory and lifestyle, already determine terms of individual and collective processes of individuation prior to social relations of property. Third, and consequently, the Marxist distinction between modes of production and social relations of production does not grasp theoretically the fact that the history of production is at the same time a technical history of memory. It is this thinking of technical memory that allows one, however, to come to terms conceptually with the latest stage of capitalism: cognitive capitalism (CPE, 30–1, 46–50/45–6, 66–72). For the latter emerges from the convergence between the objects of memory of the new technologies and the logic of the market. With this convergence, both individual memory and collective memory become increasingly subordinated to discrete terms of calculability, the rule of short-term consumerism, and lawless capitalist cynicism. The threat is, in Flaubert's terms, an age of general 'stupidity' (*bêtise*), but without the modernist relief of bourgeois art (see TC, 31/61–2; CPE, 60–6/84–92).

Stiegler reads political economy in terms of technology and the technical support. I suggest that, as a result: (1) a certain specificity of the economic is lost in the technological reading; and (2) this loss has important implications for how one conceives political adoption of contemporary capitalist forms.

In his manifesto-like *For a New Critique of Political Economy*, Stiegler rightly laments the lack of economics in recent critical thought (CPE, 16–17/28–9). 'Contemporary [critical] philosophy speaks little of the economy [. . .] as if still haunted by the supposed economism of Marxism' (CPE, 17/29). Stiegler does not look, however, to see how critical political economy may be evolving in the fields of international economics and politics in order to re-address the former from within critical thought. Rather, he emphasises that the correct response to this absence in philosophical discourse consists in re-inventing Marx's critique of political economy through technology and a constitutive understanding of the technical support (CPE, 11, 17, 19, 35/21, 28, 31, 52). Despite the interest of Stiegler's approach, this response avoids the important objects of political economy. Let me show, schematically, how and why.

Stiegler holds to the Marxist theory of the tendential decline in the rate of profit. Marx's argument is well-known (see Marx 1976, Part 7, Chapter 25: 'The General Law of Capitalist Accumulation'). With an increase in fixed capital (technology), there is an increase in productivity (more units of x good produced per hour) and, over time, fewer hours of exploitable labour. Following the labour theory of value, Marx argues that, with fewer labour-hours, there is less exploitable surplus-value. Despite counter-acting measures (industrial reserve army, capitalist expansion, mergers, etc.), the rate of profit therefore decreases tendentially, until a systemic crisis either overthrows capitalist social relations or produces a new form of them. For Stiegler, twentieth-century capital accumulation (under conditions of the decline in the rate of profit) is increasingly organised through exploitation of the consumer (CPE 27–8/41–2). It is, therefore, less the producer than the consumer who is today alienated for the sake of capital accumulation. Indeed, Stiegler provocatively suggests that contemporary consumer society is made up of a 'generalised proletariat' alienated from social determinations of life (CPE, 33/48).

Capitalism counter-acts the decline in the rate of profit by relentlessly appropriating the contemporary externalisation of (technical) memory. As a consequence, all forms of knowledge and memory become short-circuited by the contemporary 'generalisation of hypermnesic technologies' (CPE, 36/52). The nineteenth-century condition of the alienated labourer, unable to objectify himself in industrial technology, becomes that of the contemporary consumer expropriated from the possibility of digesting and synthesising at leisure the market-led networks of discrete, digitalised memory (CPE, 30–5/45–51). For Stiegler, accordingly, the present financial and economic crises confirm that decline in the rate of profit has not been overcome (as neoliberals believe) and that the intense model of consumerism based on the hypermnesic technologies has critical limits: social disorientation and de-motivation (CPE, 36–40/52–7). Against this general 'misery of spirit', technics becomes for Stiegler 'the major stake' of 'political struggle' (CPE, 36/52). Marxist-inspired critique thereby regains its practical vocation as a critical philosophy of technology. It seeks to slow down and synthesise, for individual and collective processes of individuation, the short-term horizon of contemporary technical supports. The (Greek) Platonic fear that memory is lost through its externalisation upon supports has become, in other words, generalisable under conditions of cognitive capitalism (CPE, 29/44). Much of this reconfiguration of Marx is dependent on Freudian libidinal economy. I will turn to it in the next section. First, I argue that this reading of our contemporary condition is too Greek and misses the requirements of political economy as a result. (It is also suspect from a Derridean perspective since it increasingly understands the *pharmakon* of the new technologies in ambivalent, rather than aporetic terms, but I leave this appraisal to a deconstructionist.) [. . .]

Now, for Stiegler, the question of technics is a Greek question because the relation between the human and the technical is explicitly posed by the Greeks, and any thinking on technology necessarily works *within this Greek framework*.[4] Whatever one makes of this thesis technologically speaking, the question of the modern and contemporary autonomy of the economic from the social whole is nevertheless not Greek. With the end of the Cold War, with increasing trans-border activity of capital, goods and, to a much lesser extent, labour, capital comes to determine the terms in which the allocation of scarce resources is made. Capital becomes, that is, general, and there is for the foreseeable future no alternative to it.[5] All human beings live within the system of capital, whatever the particular node they live on, or conjunction they make with it. This system is highly unstable and dissymmetrical with immense imbalances in equality, natural resource distribution, financial assets and terms of trade. With no alternative to capital, a revolutionary politics is no longer tenable. The ethical question driving political innovation has, consequently, to be worked out in terms of universally coordinated, but locally determined equilibriums between growth, sustainability and equity. Given economic interdependence and the necessity of large transfers of technology and wealth from the developed world to the developing world in the context of climate change, effective financial regulation, economic coordination and staggered development present the right strategies to tame the excesses of neoliberal global capitalism. Whether these strategies are feasible or not is at present an open question given recent government failure to regulate risk-taking and the evident dilemma, for developing countries, between the need for curtailed energy use, on the one hand, and industrialisation and exit from poverty, on the other.

Now, whatever our answers to these large questions, the political question today – 'who are we?' – can only be appraised if the political economy of a globalised world becomes the direct object of critical attention. Only by foregrounding this object and its dilemmas will one have any chance of critical purchase on the political challenges ahead. In this context, Stiegler's foregrounding of technology to promote a new critique of political economy is decisive in purpose and tone, important in detail, but misplaced in general intent. Stiegler is right to stress again the pertinence of the economy for critical thought after 'the supposed economism of Marxism' (CPE, 17/29). His technologically trained focus on the alienated consumer is important within the cognitive dimension of contemporary capitalism and debt-led growth. But, if he is concerned to show, as a philosopher, the general lines of a re-invented critical political economy, his object and attention need to be much larger than his 'Greek' framework affords. Since there is no systemic alternative to capitalism at this moment in history, the question of political economy is one of whether effective regulation of capitalism is possible or not for the world as a whole.

In this regard, I fear that Stiegler's rhetorical logic of excess testifies to a straightforward shift of Marxist terminology (from producer to consumer) rather than a reinvention of Marxism's object (political economy). [. . .] To take a few examples from only the last pages of *For a New Critique of Political Economy*: we are witnessing the 'extreme disenchantment' of the world (CPE, 59/83), a 'generalised proletarianisation [of consumption]' (CPE, 56/79), the 'disappearance of the middle classes' (CPE, 63/89), the 'destruction' of social association (CPE, 62/87), and 'lawless and faithless' elites of capitalism (CPE, 63/88). This logic of excess ignores the need today to make small distinctions, under the canopy of political regulation, within the world as a whole. The art of politics today is the prudential art of making critical distinctions within an economy of the same. 'Critical philosophy' may wish to eschew such distinctions, but it does so at its practical peril when there is no alternative to capitalism, and when, just as importantly, the mid-term horizon is global coordination of a world economy under circumstances of economic imbalance, energy crisis, and poverty.

The political questions today are therefore: 'what kind of regulation of capitalism is ethically and empirically appropriate?'; 'at what level is it appropriate?'; and 'what instance should and can decide?' These are vast and difficult questions for philosophy, political science and economics. It is my belief that, within these questions and their distinctions, an engaged philosophy (which Stiegler rightly advocates) has an important role to play. A generalised technological reading of Marx creates in this context important cultural work; but it does not give itself the terms of a contemporary critique of political economy. [. . .][6]

Economic alienation from social life should consequently not be thought within the 'Greek' framework of technology (however differentiated this framework is). Dis-embedded global capitalism requires a new international political theory of legitimate and effective regulation. The above economic alienation includes the convergence between consumerism and the logic of the market and the importance of adopting the new media and informational economies. Of these Stiegler speaks with originality and impressive intellectual force. However, technical supports – and their lack of present political adoption – do not fundamentally determine our lack of a 'we'. To argue so runs the risk of unilateral technological determinism. And this form of determinism ends up, ironically, missing its political end.

Stiegler and Freud: sublimation and de-Sublimation

In a move that has become a trait of critical French philosophy, Bernard Stiegler turns to Freudian libidinal economy to underpin his analysis of contemporary capitalism, specifically the displacement of 'alienation' from the producer to the consumer. As we saw above in his general re-reading of Marx, cognitive capitalism distinguishes itself from previous capitalist forms through the convergence between objects of the mind and the short-term logic of the market. This convergence creates the general crisis of memory and poverty of 'spirit' that marks our time. As is now clear, the convergence and its consequences call for a critico-technological response, which Stiegler advances through his re-writing of the German and French phenomenological traditions in the contemporary context of the new media (the hyperindustrial support). Through this convergence, capitalism's capture of energy for production and consumption becomes increasingly invasive and unilateral. Since human memory lies in the technical support, and since we temporalise ourselves from out of this support, our contemporary industrial condition affects the whole mind–body complex of the human (Stiegler's current term for this is 'organology'). Given, however, the rules of capital accumulation, decline in the rate of profit, and short-term profit-motivation, cognitive capitalism so captures the energy of the consumer that it blocks the sublimating processes of energy that constitute, for depth psychology, the condition of work, art, family, love and the social bond in general. Hence the importance of Freud to Stiegler, but, equally, the need to inscribe Freud's meta-psychological model of 'ego–superego–id' within the technical history of tertiary memory, retention and protention.

For Stiegler, cognitive capitalism increasingly reduces desire to its constituent drives. Stiegler calls this reduction 'negative sublimation' (DD2, 115–20, 123–4/163–9, 173–4). It implies the break-up of desire into its constituent elements of aggregation (the principle of life) and destruction (the death drive). This is a complex step in his technological critique of capitalism, and I do not have space here to develop it in full. I am also unsure that I could do so without a much deeper rehearsal of the Freudian problematic. Suffice it to say the following for my own argument. I refer to *For a New Critique of Political Economy* and *Taking Care*. Under the negentropic logic of capitalist profit and its use of the contemporary technologies, it is the young consumer who is targeted. Due to this targeting, s/he is losing her/his primary identifications. Hypermnesic technical supports (from television through CDs to the internet, all soon in the one support of the 'mobile' phone) confuse generational roles and differences and are gradually replacing the 'care' of parenthood, and its attendant authority and role-modelling, with a violent disorder of dispersed identifications without meaning or rhythm. This replacement and confusion is leading – among the younger generations that temporalise out of the tertiary memory of the new technologies – to disintegration of the family intergenerational model, disaffection and disindividuation. These generations lack – in depth psychological terms – a structuring superego to determine in their psychological apparatus the reality principle and conscience and, thereby, open up a human understanding of, and path to, law and justice. In other words, for Stiegler, cognitive capitalism attempts 'to control the id' (TC, 9/25) and 'displace primary identifications with our ascendants' (TC, 9, 13, 42/25, 31, 83) towards a new libidinal economy of commodity fetishism.

The human apparatus (technico-psychological from the beginning given ephiphylogenesis) is fundamentally threatened. Now, for Stiegler, this threat is radically new within the technical history of memory. The legacy of transgenerational identifications between the unconscious and conscious (mediated by our superego) is today undermined by the specifically psychological nature of the new 'psychotechnologies'. As Stiegler neatly puts it, Foucauldian 'biopower' is being supplemented by this new 'psychopower' of normalisation (TC, 13/31). With this replacement of the superego, desire is broken down into its primary constituents: the principle of life, on the one hand, and the death drive, on the other. Stiegler's 'negative sublimation' focuses on the death drive: the 'psychopower' of the new technologies 'destroys desire' and, with 'the confusion of generations' leads to 'nihilism' (TC, 20,

35, 41/47, 69, 79; compare also DD2, 41/65–6: 'the *reversal* of the order of generations' coincides with 'the revelation of the *economic vanity* of the destruction of [. . .] hopeless existences'). Without the primary identifications with their parents, the young generations are also unable to envisage change, since such identifications create the psychical framework within which we can alter our identities (TC, 61–2/117). When the technical support becomes hyper-industrial, and cognitive capitalism comes to conjoin mind and matter, the concomitant displacements of desire risk, in other words, the very decomposition of desire. As a result, desire no longer projects itself out as the fulfilment of itself as a non-existent justice (to come).[7] Due to this technological invasion of the id, 'public reason' radically regresses (TC, 20/47).

What is required in response, for Stiegler, is a politics of adoption of our new technological 'environment' that is centered on a re-founding of public education. The political struggle against cognitive capitalism becomes one of educating youth so that the young can begin to synthesise the deep attention span of learning and critical reflection with the market-led hyper-attention of zapping (TC, 73–9/137–43). This politics of education would prevent present technological sophists from destroying the legacy of 'spirit' (unconscious and conscious retention/protention), return the technological *pharmakon* to its proper ambivalence, and open up the future. As Stiegler rightly argues in the context of the internet, we need to 'envisage new processes of transindividuation' (TC, 86/158).

This overall argument on 'psychopower' is dynamic, imaginative and singular in its use of the 'French' legacy of Freud. I will briefly pose some questions that rhyme with my larger observations concerning his re-writing of Marx. The question is again one of the specific autonomy of the domain under technological consideration and the political consequences of losing this autonomy theoretically.

Just as Stiegler gives us a technological reading of political economy, so he also gives a technological reading of libidinal economy. Technics constitutes the condition of sexuality *qua* desire. This critique of Freud inscribes the whole of the psychical apparatus within the technical history of epiphylogenesis. It is clear that human sexuality has both evolved and is altered through technical developments. Stiegler is right to insist, with the paleontologist Leroi-Gourhan and Gilles Simondon, that hominisation is a technical process of evolution and psychic and collective individuation. That said, sexuality is not reducible to technics. Human sexuality, together with the problematic of desire that it underpins, both transcends technological determination and is itself dependent on many variables. There are psychoanalytic constants (for example, the Oedipus complex) that determine the transgenerational legacy of the id beyond technical evolution. To argue otherwise (as Stiegler does – see 1996) is not to engage with the autonomy of the depth psychological. What with the neurosciences' penetration into the mind–body complex, we are probably only now beginning to understand this autonomy and multi-causality. Stiegler is therefore correct, following Herbert Marcuse, to place technics within the evolution of sexuality and the vagaries of desire. There would be no Oedipus complex, specific to human animals, without the technological evolution of the human. But he goes too far when he makes the relation between technics and desire one of unilateral determination. The above argument that the 'psychotechnologies' are attempting 'to control the id', if not 'the psychical apparatus in general' (TC, 13/31), is one consequence of this unilateral determination. This is another technologically determinist judgment. It makes a background condition (technology) into a radical determination of the psychic apparatus as a whole. Such determinism tempts Stiegler into arguing for a general 'crisis of spirit' at the moment of cognitive capitalism.

Let me recall in this context that, for Freud, sublimation (the turning of desire into law) constitutes a complex process that is dependent on many contingent factors. In distinction to all other animals, humans sublimate because they are diphasic: we undergo the latency period (and therefore puberty) on account of our technological specificity. As a result of this diphasic nature, the human animal turns its love of its protectors into an identification that, with the reversals of puberty, comes to structure and occupy the space of the superego. Identifying with our parents (and their parents, etc.) or taking distance from them constitutes, from the beginning, a complex process of love and hate that may lead, from puberty onwards, to too rigid a superego or too dissipated a one (or rather, to variations in between). Freudian psychoanalysis suggests that it is very difficult to generalise with regard to this development. The absence of identifiable, recurrent and protecting love can indeed create an uncoordinated psyche.

It leads, in this case, to other forms of parental identification that are always ongoing in the infantile years precisely because the id transcends technically organised memory. Until the nuclear family is literally dissolved and not replaced by another form of social organisation, we cannot consequently speak of a new generation that has lost its primary identifications and, therefore, following the Freudian logic of sublimation, lost a sense of the future, of law and of justice. There are too many variables at play within the psychoanalytic dynamic of infantile protection and care for Stiegler to be so clear. Under new conditions of technology, one must be proactive and prudently regulate internet flows (regarding collective security, obscenity, etc.). One must, however, wait to see what new forms of parenthood adopt the hyperindustrial support and what new forms of sublimation will come to structure the coming generations' sense of conscience. These new forms may be weaker than either traditional or modern forms of the close social bond. But this cannot be a cause of excessive concern – unless this polemical pitch is judged to be the right means to attract political concern and change public policy (and even here, I am unsure that it is). Ontologically speaking, these forms may lead to more innovative and creative behaviour as much as to destructive and self-destructive behaviour. I am arguing that we cannot know at this very early stage of our hyperindustrial age, although Stiegler is nevertheless right to call for critical synthesis. The political adoption of the hyperindustrial support will take time – as did monotheism to adopt non-orthographic writing and the social contract to adopt the alphabetical word.

The above uncertainty regarding the direction of the contemporary technology–human symbiosis is in Stiegler's terms the 'ambivalence' of technology. In Freudian terms, it is more simply the complexity of the human mind–body complex (on these themes, see Beardsworth 1996b). In these processes there is a constant dialectic between 'negative' and 'positive' sublimation: here, the reduction of law to capitalism on the one hand, and the embedding of capitalism within artistic and legal forms on the other. Stiegler cuts the knot of this ambivalence too quickly, or rather, generalises too fast from exclusively French examples of de-sublimation (see *Taking Care*, especially on the advertising techniques of Canal J.).

Regarding Freud, I would argue, in sum, that Stiegler gives a strong, original reading of contemporary affective life through the bridging of technology and the psyche. Conversely, it is a techno-logical re-reading of Freud that flattens out the vagaries of human affect and human conscience, preventing a nuanced, comparative account of the relation between contemporary consumerism and normative thought and behaviour. As a result, public educa-tion may be posited too quickly by Stiegler as the right political response.

This is not to deny the need for change in public education: far from it. The internet clearly poses a problem. As the contemporary teacher knows, internet-surfing produces a form of consciousness that is adept at 'copy and paste', but finds synthesis and judgment increasingly difficult. Stiegler's politics of critical reflection, with its emphasis on the vital role of education, is in this sense persua-sive. That said, I would wish to keep a sense of global perspective. As is well known, use of the internet was crucial to the election of Barack Obama: it helped create a cultural transformation that proved strong enough to shift the American political landscape to the center. The use of mobile phones has transformed the electoral process in West Africa. The internet is, in other words, already highly creative politically.[8] Education must certainly help to supplement this emerging creativity with the art of judgment. Obama's domestic fate regarding healthcare reform since the campaign has shown, at the same time, how powerful the tradi-tional media remain in shaping political perception and interests. Progressive liberal politics in one of the most technologically savvy of countries depends as much today on restructuring the power-bases of the traditional media as it does on providing an educa-tion in response to capitalist-led technological transformation of human memory. Stiegler would not disagree with this last point. As I said at the beginning of this paper, his political voluntarism was in the 1990s singular on the French theoretical continent. It means, to my mind, however, that philosophico-political reflection should consider the political adoption of technology at several levels of analysis and of policy, and in a spirit of prudence and intellectual modesty.

Conclusion

I have addressed the work of Stiegler through the names of Marx and Freud. In doing so, I have attempted to suggest that his work goes too quickly over what puts a brake on the destructive side to capitalism. The future political project of democracy is, without doubt, to embed capitalism at the world level. And democratic freedom means that one must renounce gratifying one's immediate desires. This means political institution and self-restraint. Stiegler focuses rightly, and sometimes brilliantly, on the urgency of the political today, and on the importance of a political adoption of contemporary forms of industry within a general intellectual framework of retentional finitude. Not to analyse the forms of institutional change at the appropriate level and not to give credit to the specificities of sublimation within capitalism tend, however, to make capitalism's field and dynamic too uniform, and Stiegler's responses to it too unilateral and too general (if not too French). As a result, his theoretical world turns too quickly, at the precise moment when a slower speed and a finer set of distinctions are needed. Not that there is not enormous danger in our present world, not that a sense of urgency is not vital. Our description of it requires, however, theoretical terms that exposit it in its complexity so that theory can provide, precisely, the occasion for suitable political adoption and decision.

Notes

1. *Editors' note*: An earlier, longer version of this piece was published in *Cultural Politics*, 6.2 (2010): 181–99. Notable excisions from this original have been marked with an ellipsis ([. . .]) in the text and, on occasion, are flagged by a footnote indicating the gist of what has been removed.

2. See particularly *Technics and Time, 2: Disorientation*, Chapters 2 and 3, 'Genesis of Disorientation' (TT2, 65–96/83–116) and 'The Industrialisation of Memory' (TT2, 97–187/119–216).

3. The most virulent critique of global capitalism in these terms is to be found in Stiegler DD3.

4. This is one of the major theses of *Technics and Time* as a whole, and is formulated in Volume 1.

5. The problems of the planet's ecosystem do not pose, for example, a radical challenge to the system of capitalism (allocation of resources under the principles of capital accumulation) at this stage of history. They demand major re-inflections, ones which the market institution, together with appropriate regulatory bodies at public and private levels, will adapt to and put in place (carbon cap-and-trade and clean development mechanisms being the two major innovations that await

global dissemination). Stiegler's work on global capitalism has not focused on them, however.

6. EN: At this point, the original article moves into a discussion of the 2007–9 financial crisis, of which Beardsworth states: 'I remain unconvinced that Stiegler's philosophico-technical reading of the economy can (1) properly delimit the economic problems that need to be adopted; and (2) tease out the differences of approach required to adopt contemporary economic conditions effectively' (Beardsworth 2010: 190–1).

7. After recasting the Derridean trace in terms of tertiary retention, Stiegler here re-amalgamates his theory of technology with the Derridean and Levinasian themes of radical alterity *qua* justice. The re-amalgamation constitutes a powerful theoretical move based on the Lacanian problematic of desire. It lies at odds, however, with Stiegler's initial concern to give a material history to the trace through the concept of tertiary memory – unless Stiegler makes more distinction within his own philosophy between the hyperethical and political adoption.

8. Although the comment on Barack Obama in 2009 was too optimistic, that the new social media have served as a catalyst to the social uprisings in the Middle East shows how politically dynamic the internet already is.

14

Memories of Inauthenticity: Stiegler and the Lost Spirit of Capitalism

Ben Roberts

The aim of this chapter is to discuss Bernard Stiegler's arguments about the condition and fate of contemporary capitalism. These arguments are formulated most succinctly in *For a New Critique of Political Economy*, but elements of this 'new critique' are discussed at greater length throughout Stiegler's recent work and in particular in the series *Mécréance et Discrédit* (*Disbelief and Discredit*) and *De la misère symbolique* (*Symbolic Misery*). In the following discussion I have chosen to focus primarily on the third volume of the former, entitled *L'Esprit perdu du capitalisme* (*The Lost Spirit of Capitalism*), the first part of which is dedicated to a detailed critique of Luc Boltanski and Ève Chiapello's *The New Spirit of Capitalism*. There are two reasons for this focus. The first is that Boltanski and Chiapello's book has become an extremely influential account of contemporary capitalism; Stiegler's arguments here help to demonstrate the problems and limitations of its approach. The second is that the contrast with Boltanski and Chiapello helps to bring out much that is genuinely novel and interesting about Stiegler's own account of capitalism. It demonstrates in some detail the way in which his transformation of work by Derrida, Simondon and others has allowed a rearticulation of some of the concerns of the Frankfurt School. It also shows how in his later work Freudian concepts have extended his diagnosis of 'disindividuation' in contemporary capitalism.

Boltanski and Chiapello distinguish the 'social critique' of capitalism, embodied in a demand for equality, from its 'artist critique', which they associate with a demand for liberation and authenticity. They go on to argue, based on a comparison of 1960s and 1990s French management literature, that contemporary capitalism can be understood partly as a response to the artistic critique, which it attempts to assimilate and incorporate. In this essay I will discuss Bernard Stiegler's criticisms of their work, including his objection that we need to understand current developments in terms not of a new but of a 'lost' spirit of capitalism. I will go on to consider what the implications of this debate are for his wider project and in particular its critique of 'hyperindustrial' capitalism.

The new spirit of capitalism

To prepare the ground for this discussion we need to recap briefly the aims of *The New Spirit of Capitalism*. Boltanski and Chiapello in fact distance themselves from the project that the title would seem to suggest they are undertaking, that is, a contemporary reworking of Weber. They put to one side the wider question of the ethos of capitalism, preferring to define spirit more narrowly as 'expressed in a certainty imparted to cadres about the "right" actions to be performed to make a profit, and the legitimacy of these actions' (Boltanski and Chiapello 2005: 16). As they put it:

> We shall set aside the predispositions towards the world required to participate in capitalism as a cosmos – means-end compatibility, practical rationality, aptitude for calculation, autonomisation of economic activities, an instrumental relation to nature, and so on – as well as the more general justifications of capitalism produced in the main by economic science [. . .] *Our intention is to study observed variations, not to offer an exhaustive description of all the constituents of the spirit of capitalism.* This will lead us to detach the category of spirit of capitalism from the substantial content, in terms of ethos, which it is bound up with in Weber. (Boltanski and Chiapello 2005: 11 [original emphasis])

The key point here is the focus on 'observed variations', which prepares the ground for the sociological empirical work on which the book's claims rest. The wider question of 'ethos' is therefore not addressed by *The New Spirit of Capitalism*. If the concept of spirit here is narrower than in Weber, it is perhaps partly in order to justify the focus on different eras of management discourse. Boltanski and Chiapello's spirit of capitalism provides those engaged in managing capitalism (but not its main beneficiaries) with an enthusiasm for their endeavour, a minimum of security and some sense that their activity is compatible with the common good (2005: 16).The focus on comparison and variation also indicates that this is not so much a thesis about capitalism in general as about changes *within* capitalism, and particularly those changes in capitalist spirit that the authors identify between the 1960s and the present day.

Indeed, Boltanski and Chiapello identify three distinct spirits of capitalism: the first, which has its roots in the nineteenth century and can be seen in the figure of the bourgeois entrepreneur and the family business, stresses 'gambling, speculation, risk, innovation' (2005: 17). The second, which develops between the 1930s and 1960s, is the spirit of the bureaucratic, centralised large corporation and emphasises economies of scale, production standardisation [and] the rational organisation of work' (2005: 18). The third and eponymous 'new spirit' of capitalism has emerged between the 1960s and the present day and consists in a highly decentralised networked form of capitalism, characterised by 'flatter' organisational hierarchy, much greater autonomy within firms for both individuals and teams, lower job security and the proliferation of temporary contracts and outsourcing.

Each spirit of capitalism also incorporates a *justification of capitalism* (2005: 20) and the changes between different spirits can therefore be understood as shifts in the nature of this justification in response to the various criticisms that are made of it. The evolution of the spirit of capitalism is therefore intimately bound up with the criticisms that are made of the capitalist mode of accumulation. Chiapello categorises these as follows:

> Capitalism is criticised: (a) as a source of disenchantment and of inauthentic goods, persons, and lifestyles; (b) as a source of oppression that is opposed to freedom, autonomy, and creativity; (c) as a source of misery and inequality; and (d) as a source of egotism entailing the destruction of forms of solidarity. (Chiapello 2004: 586)

For Boltanski and Chiapello – and this is where their analysis itself moves into 'authentically' new terrain – the first two criticisms belong to something which they propose to call artist or artistic[1] critique of capitalism, the latter two belong to its 'social critique'. In other words, artist critique is generally characterised by demands for autonomy and authenticity, whereas social critique opposes inequality and egotism. These two critiques, though sometimes exemplarily combined – as for Chiapello is the case with the Frankfurt School (2004: 586 n. 5) – are not straightforwardly compatible and may even conflict, typically being articulated by distinctively different social groups. The rather more substantial claim here is that the 'new spirit' of capitalism, as evinced by their comparison of management literature from the 1960s and 1990s, is partly a response to the artist critique that they identify, particularly as embodied in the movement culminating in May 1968.[2] That is to say that 1990s-style 'neo-management aims to respond to demands for authenticity and freedom, which have historically been articulated in interrelated fashion by what we have called the "artistic critique"' (2005: 97). This response to artist critique consists not only in less bureaucratic, 'more human', forms of organisation but also the invention and marketing of products which are 'attuned to demand, personalised, and which satisfy "genuine needs"' (2005: 99).

Stiegler's longest sustained account of Boltanski and Chiapello is in *L'Esprit perdu du capitalisme* (*The Lost Spirit of Capitalism*), the third volume of his series *Mécréance et Discrédit* (*Disbelief and Discredit*). As the title suggests, while endorsing some aspects of their work, Stiegler sharply disagrees with their diagnosis concerning the 'spirit' of contemporary capitalism. The current situation is one in which capitalism is in crisis and that crisis is characterised by a loss of spirit:

> Although I often agree with these analyses, I don't, for all that, believe that there is nowadays a 'new' spirit of capitalism, whatever that may be – nor that this would be its third stage. On the contrary I believe that what is at stake and in question here is the *libidinal economy of capitalism, in as much as it leads to the liquidation of all the sublimations that constitute the superego, which is always supported by processes of sublimation, the fruits of which we call 'spiritual works'.* (DD3, 18 [original emphasis])

As can be seen in this quote, Stiegler's understanding of 'spirit' is rather different from the carefully constrained sense in which, as we have seen, Boltanski and Chiapello use it. In one way he stays closer to Weber, seeing in contemporary developments a generalisation of the 'disenchantment' that was described in the latter's *The Protestant Ethic and the Spirit of Capitalism* (Weber 1958; DD3, 19). However, Stiegler also redefines this loss of spirit here in Freudian terms, as desublimation and as destruction of the libidinal economy on which capitalism depends. Allied to this Freudian analysis is an understanding of spirit as psychic and collective individuation, a term drawn from the work of Gilbert Simondon (2007). In effect the analysis of 'loss of spirit' is a blend of desublimation and disindividuation. As Stiegler puts it, 'the process of psychic and collective individuation – in which the simultaneously conjunctive and disjunctive *and* is, I believe, constitutive of spirit itself [. . .] – presupposes sublimation' (DD3, 29). Stiegler's argument here about loss of spirit therefore recalls earlier arguments about the loss of individuation drawn from his reading of Simondon.

In *Du mode d'existence des objets techniques* (*On the Mode of Existence of Technical Objects,* 1958), Simondon argues that the rise of the machine tool removes the ability of the skilled worker to differentiate his or her labour from that of other workers (Simondon 1989: 77–82, 118–19), in what Stiegler calls a 'loss of individuation' that is exacerbated by the new teletechnologies and their industrialisation of memory (DD1, 57). As he puts it in the first volume of *Symbolic Misery,* 'the current loss of individuation is a stage of grammatisation in which three individuations, psychic, collective and techno-machinic, generalise formalisation through calculation' (SM1, 142). Here Stiegler adds to Simondon's analysis the idea that the process of industrialisation is also a *grammatisation,* that is to say, a process, analogous to that of the development of writing, by which idiomatic actions (for example, those of the weaver) are standardised, discretised and materialised (for example, in the Jacquard loom, CPE, 10–11):

> Industrial machines allow the actions of manufacture to be discretised, formalised and reproduced, duplicating the know-how of the worker and putting it into the machine. Industrial machinery reproduces the actions of labour in the same way that writing permits the reproduction of speech and printing allows the reproduction of copies. It also permits the rationalisation of actions, just as writing allows the rationalisation of a particular way of writing – and therefore of speaking, of thinking – in what will become a homogeneous linguistic milieu. (EHP, 57)

This stress on industrialisation as grammatisation is an important distinguishing feature in Stiegler's account of contemporary capitalism. He does not accept the thesis that modern societies are 'post-industrial'. Rather he regards them as hyperindustrial, as the continuation of modernity towards 'the industrialisation of everything' (DD1, 102–3, 109–10; SM1, 98). The hyperindustrial marks the extension of industrialisation from the mechanical to the computational, from the manufacture of goods to the industrialisation of memory and culture, and a corresponding expansion of proletarianisation from the producer to the consumer, from the working to the middle class (DD1, 62). This is because grammatisation is at the heart of the way in which capitalism is to be understood, that is, capitalism is itself a specific epoch in the Western process of grammatisation:

> Beyond that 'rationalisation' described by Weber, and the separation of capital and labour described by Marx, capitalism is the expression of a tendency towards the *mechanical externalisation* of that which characterises the singularities composing the process of individuation; and, as such, it is the *mechanised epoch* of what in *De la misère symbolique, 1,* I have called *grammatisation*. Nevertheless this tendency, as mechanical exteriorisation, has the effect of producing a standardisation and a formalisation, submitting everything that it formalises to *calculability*. (DD1, 38)

What is unique about contemporary capitalism in its current epoch, then, is that this tendency toward exteriorisation has become hegemonic (DD3, 37), resulting in the hypersynchronisation of 'singularities' and the loss of individuation or disindividuation which we have just discussed. Unless singularities can be 'recast [*rejouer*] on a new plane', this tendency can lead to disaster; in other words, we need to avoid the risk that 'capitalism could *brutally* and *prematurely* come to a very bad end, and us with it, though the *decomposition* of its tendencies' (DD3, 27; DD1, 38); Stiegler's concern here is to save capitalism from itself.

There are a number of obvious differences here with the analysis provided by Boltanski and Chiapello. The standardisation that they associate particularly with the 'second spirit' of capitalism (roughly 1930s–60s), for Stiegler forms part of an essential tendency towards mechanical exteriorisation and is pushed even further in today's creative or cognitive capitalism. For Boltanski and Chiapello (writing in 1999, so before the global financial crises of 2007–8), capitalism is 'flourishing' (2005: xxxvi); for Stiegler it is on the verge of self-destruction (DD3, 26–7). What they see as 'ideological changes' within capitalism, he sees as different expressions of a continuous tendency (to exteriorisation). Indeed, beyond the grammatisation thesis, Stiegler's understanding of the limits of capitalism are classically 'tendentialist': a tendency for the rate of profit to fall, a tendency for the libidinal energy of consumption to become exhausted, a tendency to exceed the natural resources of the planet (CPE, 23–5, 92).[3] In other words, he sees capitalism as transformed through its own inherent dynamics, through reaching its tendential limits.

On the other hand, the authors of *The New Spirit of Capitalism* depict the spirit of capitalism as evolving in response to the critiques made of it, as a result of the 'three-sided game' between critique, capitalism and its legitimising spirit (2005: 29–30). The emphasis here is on the sociology of management cadres and very short shrift is given to economic science as a means of understanding capitalist spirit. Boltanski and Chiapello are equally dismissive of the idea that changes can be understood in technological terms, seeing here the risk of 'technological determinism' (2005: xix). Therefore, despite their emphasis on the centrality of the network form to the new spirit of capitalism, there is very little discussion of information networks. As Stiegler argues:

> What is very much neglected in the analysis of Luc Boltanski and Ève Chiapello, in other words, is not only the place of the media, but also that of technics, which is presented here only as a connexionist model, the implications of which need to be verified anyway (why connexion rather than rhizome, for example?) What they frequently evoke as personalisation and *one-to-one*, 'or *point to point*' contact, is a result of network connectivity and technical evolution long before being the response to so-called 'demand' [. . .] The role of information and its technologies is not analysed at all. (DD3, 53–4)

The argument here is that the 'connexionist' model is not simply, as Boltanski and Chiapello argue, a change brought about by artist critique of captitalism's conformity, formulated as a demand for 'personalisation'. Rather it is underpinned by a technical and informational evolution that begins long before the changes in management practice associated with the 'new spirit'. To assert, as Stiegler does here, the importance of these technological supports is not necessarily to argue for a form of technological determinism, or the assumption that technological change drives social change in a deterministic fashion. As I have argued elsewhere, Stiegler's ideas around technics and grammatisation do not in fact privilege technology as the determining factor in social relations (Roberts 2012). This is partly because he sees technics, as the exteriorisation of memory, as itself a form of contingency, that is, an escape from the biological programme. But it is also because from the beginning of his work Stiegler describes the prosthetic relationship between the human and technics as a structure without origin, where neither side plays the role of 'cause' to the other as 'effect' (this is analogous to the way Derrida sees language as *différance*, as a structure without origin). In such a structure to talk of a determining relationship, such as technological determinism, makes no sense. This is why Stiegler talks of the 'differance of the human' in the first volume of *Technics and Time*, for example (TT1, 134–45). But he also makes this argument in relation to the centrality of spirit in Boltanski and Chiapello's account:

> It is not ideas that cause phenomena. Which is not to say that it is *something else* that *causes* ideas. It does not signify, for example, that technical individuation or even transformation of the relations of production would be the cause of the 'superstructural' creation of ideas. (DD3, 49)

In other words, to suggest that capitalism's new networked spirit cannot be thought separately from the materiality of information networks themselves is not an argument that changes in technology can simply explain the transformation of spirit. Rather it is to see the two in a relationship of 'co-individuation' within a single process:

> Are networks the origins of organisations and ideas, or the inverse? *False problems*. Nothing is the origin of anything else: all these – individuals, ideas, organisations, technics – are formed in a process. (DD3, 50 [original emphasis])

The problem for Stiegler, then, with Boltanski and Chiapello's emphasis on the role of spirit is that it leaves in place a virtually uninterrogated dualism of materialism/idealism or base/superstructure (49–50). But it is just this opposition between base and superstructure that the thinking of technics, and in particular *mnemotechnics*, the technical preservation of memory, refuses.

As Stiegler puts it, 'retentional supports, infrastructure, constitute retentional apparatus, superstructure' (DD3, 57). As a result, the opposition between infrastructure and superstructure is 'inoperative' (DD3, 57).

Consumption and the role of authenticity

Another facet of the narrow way in which Boltanski and Chiapello define 'spirit', or the separation of spirit from ethos which we discussed above, is that their focus on management cadres is also a focus on production and a relative neglect of the role of consumption. This leads to an emphasis on the management rather than *creation* of consumer demand. Indeed at times it can seem as if demand is almost naturalised within the account. Stiegler remarks on this tendency in the following passage from the discussion of 1990s management discourse in *The New Spirit of Capitalism*:

> New-style management does indeed offer various responses to the critique of disenchantment by promoting the creation of products that are attuned to demand, personalised and which satisfy 'genuine needs', as well as more personal, more human forms of organisation. (Boltanski and Chiapello 2005: 99)

But, as Stiegler puts it, 'is there really a demand?' (DD3, 38) And what is meant here by 'genuine need', even when surrounded by quotation marks? For Stiegler, such a social demand is always created and managed by capitalism, 'only an artifice in the service of generalised proletarianisation' (DD3, 39). The apparently more 'personalised' response to this demand simply disguises more addictive forms of consumption. Indeed personalisation is but one more step in 'calculability applied to the control and reduction of singularities, here reduced to particularities' (DD3, 39). (This emphasises again the continuity between the industrial and 'hyperindustrial'.)

The relatively limited role played by consumption in *The New Spirit of Capitalism*'s account is linked to a neglect of the role of desire in capitalism. Stiegler notes that the term is only really discussed in relation to Marx near the end of the book, where the idea that capitalism must create consumer desire is ascribed to artist critique (2005: 427). Boltanski and Chiapello therefore mostly ignore the ways in which capitalism does not merely respond to

desire indirectly through the action of critique, but is itself constituted through the creation, management and channelling of desire.

Stiegler's arguments here could be seen as evidence that what
his analysis offers is a form of what, as we have seen, Boltanski
and Chiapello call 'artist' critique. His revival of psychoanalytic
concepts to describe capitalism's 'loss of spirit' and reworking of
Marcuse in the second part of *The Lost Spirit of Capitalism* seem
to underline his affinity with the thinkers Boltanski and Chiapello
associate with the 'artist critique' of 1968. It is no accident that
Stiegler's readers have often made the comparison between his
work and that of the Frankfurt school (Sinnerbrink 2009; Roberts
2006). Indeed superficially his diagnosis of a 'loss of individuation' in 'hyperindustrial' societies seems easily assimilated to what
Boltanski and Chiapello label, in a problematically loose fashion,
'the critique of inauthenticity'.

As we have seen, the critique of inauthenticity is at the heart of
what Boltanski and Chiapello think of as 'artist critique'. But it is
not always clear what they mean by this term. As Stiegler puts it:

> One asks oneself if the word comes from Heidegger, from situation-
> ism, from Marcuse or elsewhere. *None* of these sources truly refer
> to this term, not even Adorno: it is rather the journalistic and leftist
> vulgate who use this very poor jargon. (DD3, 39)

As this suggests, *The New Spirit of Capitalism* is quick to include
a wide range of philosophical positions under the umbrella of the
'critique of inauthenticity'. This becomes obvious in a passage
near the end of the book where it is argued that even though
Adorno explicitly rejects the concept of authenticity in Heidegger,
the *Dialectic of Enlightenment* is 'perfectly compatible with the
Heideggerian thematic of inauthenticity' (Boltanski and Chiapello
2005: 440–1; see also Adorno 1973, 1979). For these purposes it
appears as if the critique of 'standardisation', 'massification', 'con-
sensual levelling', 'conformist domination' are all equivalent (and
indeed subsumed under) the 'critique of inauthenticity' (2005:
441). As Stiegler points out, this is apparent in a passage which
describes how the commodification of the authentic 'presupposes
sources of authenticity [. . .] such as human beings, scenery, cafés
where people feel comfortable, tastes, rhythms, ways of being and
doing, and so on, which have not yet been introduced into the
sphere of commodity circulation' (2005: 444). Here it appears

that authenticity means little more than those aspects of human existence which have not yet been commodified. Yet this is a long way from the very technical sense in which Heidegger uses the word (Heidegger 1962: 68 ¶9). Perhaps this is partly a translation issue between French and German: as Stiegler points out, authenticity (*authenticité*) is a not unproblematic rendering of Heidegger's *Eigentlichkeit*, and the sort of thing that Boltanski and Chiapello are describing might be closer to the German *Echteit* or *Authentizität* (DD3, 40).[4]

This sweeping interpretation of authenticity becomes even more problematic when it comes to the intellectual trends which *The New Spirit of Capitalism* argues are preventing a contemporary revival of artist critique. In a short but puzzling section entitled 'The discrediting of the quest for authenticity' (2005: 453–5), its authors argue that since the 1960s, authenticity has been subject to a 'systematic labour of deconstruction' (2005: 453). The outcome of this deconstruction has apparently been to undermine the critique of the inauthenticity of capitalism. The principle thinkers responsible for this deconstruction (and who seem to be denounced here) are Bourdieu, Derrida and Deleuze. Let us leave to one side the rather strange grouping of these three thinkers, and the loose use of the term 'deconstruction' as a uniform description of their work. But to take just one example, that of Derrida: Boltanski and Chiapello's argument here seems quite tendentious. It is a bit of a stretch to argue that Derrida's deconstruction of speech and writing 'dismantles a figure that has, since Rousseau, constituted one of the most fertile springs for sustaining a demand for authenticity' (Boltanski and Chiapello 2005: 454). Perhaps realising here a need to 'show their working', the authors carry on this argument in a footnote. In response to Derrida's reading of Rousseau in *Of Grammatology* (they describe it as a 'long commentary'; it certainly is longer, and much more careful, than their own one-paragraph summary of Derrida), they argue:

In Rousseau, we doubtless find the first systematic expression of the requirement of authenticity in its modern form. The voice, as authentic presence and absolute proximity of the self to itself (and hence as truth), is opposed to writing as distance, mediation, loss of presence, and paving the way for illusion, just as the immediacy of the popular festival is contrasted with the artificiality of the theatrical spectacle, and, in another respect, with the way that direct democracy through the assembly of citizens is opposed to representative democracy, in which the general will [...] risks being diverted and degraded into particular interests. (2005: 480 n. 89)

The net of authenticity is now being cast even wider. Even the philosophical privilege of speech over writing turns out to be foundational to the critique of capitalist inauthenticity and needs to be defended from Derrida's reckless dismantling. But if Boltanski and Chiapello really want to establish the centrality of what Derrida calls phonocentrism to the thinking of authenticity, they need to do so much more carefully than this. Similarly, given that the concept of authenticity is central to artist critique and given the assertion here that Rousseau is its foundational modern thinker, it is somewhat surprising that they do not dedicate some longer discussion to his work in the main text and relegate it instead to a footnote. Part of the problem here is that *The New Spirit of Capitalism* seems to move, without indicating as much, from a sociological and *descriptive* account of authenticity as a historical element in the critique of capitalism's second spirit to an *analytic* concept that turns out to be the foundation of critique in general. But this makes it even less plausible that they dedicate so little space to discussion of writers and thinkers whom they regard as crucial to the foundation and 'dismantling' of artist critique. A book that has been described approvingly as a 'dizzying theoretical tour of the past thirty years' is perhaps sometimes *too dizzying* (Budgen 2000: 150). As Stiegler points out, just four pages (in the French edition) are dedicated to the discussion of Derrida, Deleuze and Bourdieu, in comparison to 'hundreds and hundreds' devoted to the analysis of management texts (DD3, 44). Moreover, given their general treatment of singularity and authenticity as synonymous, it is odd that they do not discuss the theme of singularity in Derrida's work, nor that of idiomaticity, which is 'at the heart of the question of writing, the trace, difference, and so on, precisely *as* the question of singularity which exceeds that of singularity' (DD3, 44–5).

Capitalism and its critique

One of the strange aspects of Boltanski and Chiapello's 'three-sided game' between capitalism, its critique and the 'spirit of capitalism' is that they focus almost exclusively on the interaction between the latter poles, and have much less to say about capitalism itself. At the outset of the thesis it is given a 'minimalist' definition as the 'imperative to unlimited accumulation of capital by formally peaceful means' (2005: 4). In general this definition seems to encompass a Marxist understanding of the division between capital and labour (2005: 6–7). But almost immediately the emphasis shifts to the question of how this 'absurd system' is legitimised through its spirit. It then becomes a question of the success or otherwise of critique in challenging and changing this spirit. Capitalism as an economic system is largely marginalised in this account. It is dealt a (presumably winning) hand in their 'three-sided game', but its actual role in the interplay is rather mysterious. Everything happens as if the real game is 'spirit-critique' rather than 'capitalism-spirit-critique'. Indeed it looks as if this is not so much a three-sided game as a dialectic. In one sense this is perhaps an understandable feature of *The New Spirit of Capitalism*'s sociological account. Presumably the fear is that too close a focus on economics leads to a neglect of the social and moral aspects of capitalism with which the book is concerned. Economic science is then dismissed as just one aspect of capitalism's legitimating spirit (and not even one that we should worry about too much) (2005: 12–14). But the problematic outcome of this is that capitalism *as a system* seems deprived of any proper history or dynamic, other than that of interminable accumulation. It is as if capitalism itself were a sort of constant force modulated through the changing ways in which it is legitimised. So factors that one might think belonged to the history of the capitalist system, such as the evolution of technology, changes in productivity, the rise of information networks, financialisation and so on, are sidelined by Boltanski and Chiapello's analysis.

Boltanski and Chiapello construct a convincing account of 1968 as a crisis in the legitimising spirit of capitalism, but the battle between spirit and critique cannot be the source of every crisis in capitalism. (It seems doubtful that the financial crisis of 2007–8 was the result of brokers, traders, quantitative analysts and risk managers at firms like Lehman Brothers spending too much time reading anti-capitalist blogs.) What is excluded in Boltanski and Chiapello's account of the changing face of capitalism is the hypothesis that the spirit of capitalism might also respond to crises in capitalism itself, and not just as the result of artist and social critique. Suggesting that capitalism as a mode of accumulation might have its own dynamic and history, *including its relationship with technical exteriorisation*, does not necessarily mean either endorsing technological determinism or setting up economics as the final determining instance. It is precisely here that Stiegler's more nuanced account of the relationship between technics and culture comes to the fore – because it allows us to see the importance of the material specificity of technologies without simply opposing this dynamic to the human or contingency in general.

What is interesting is that, despite having highly divergent diagnoses concerning contemporary capitalism, both the authors of *The New Spirit of Capitalism* and Stiegler agree on one very important point, which is the need to renew critique. Indeed, in a chapter entitled 'The Test of the Artistic Critique', Boltanski and Chiapello argue that artist critique itself needs to recover both from the unfortunate outcomes of its own success (that is, the incorporation of aspects of the critique, such as worker autonomy, in the 'new spirit' of capitalism) and, of course, as we have seen, from the 'disturbing effects' of the 'discrediting' of authenticity by Derrida and others (2005: 419–82). However, their ideas on how this might be realised (beyond some legal codification of worker autonomy) seem to be mainly constrained to resisting commodification of new spheres of human activity (2005: 470–2). On the face of it this does not seem to add anything to artist critique that was not already present. It seems to me that Stiegler, on the other hand, has some fairly clear ideas about how artist critique (not under that term, of course) might be revived and rethought.

To understand how this is so let us return to Boltanski and Chiapello's polemical account of how Derrida undermines the critique of capitalism by undermining one of the most important foundations of authenticity, the priority of speech over writing in Rousseau. As we have seen, the authors of *The New Spirit of Capitalism* argue that Rousseau founds authenticity and founds it on a metaphysics of presence. Any attempt to deconstruct such a metaphysics thereby derails the critique of capitalist inauthenticity. But this assumes that the critique of inauthenticity requires the positive affirmation of the 'authentic' and, beyond that, metaphysics. In a sense (that is, in the very general one in which Boltanski and Chiapello use the term 'authentic') it is exactly this impasse that Stiegler's account of individuation seeks to avoid. Criticising the 'loss of individuation' in hyperindustrial capitalism does not imply, for Stiegler, the affirmation of an authentic individual, present to her or himself. Rather the *process* of individuation is conceived upon Derridean lines as a differing deferral of presence that is disrupted or short-circuited by capitalism in its 'hyperindustrial' current form. At no point does this mean that we found either the individual or collective group (people) on the basis of a self-sufficient 'presence'. That is the major point of Stiegler's deployment of the Simondonian concept of individuation. It is not about affirming the voice of the individual or people against the impersonal forces of technological capitalism. It is instead about the 'ecological' preservation of the process of individuation in which these are formed, one which is implicitly technical (and therefore also implicitly concerned with difference in Derrida's sense). Indeed one might argue that one of Stiegler's principal achievements is to see how the deconstruction of speech and writing (understood as an argument about fundamental technicity, or default of origin) could be rethought precisely *as* a critique of the inauthenticity of capitalism. Of course, he does not conceive it in quite those terms, not being committed to Boltanski and Chiapello's terminology or, more widely, to the conceptual baggage of authenticity. But undoubtedly his arguments about disindividuation and symbolic misery do fall under their general rubric of 'artist critique', that is to say, the critique of the tendency of capitalism towards homogeneity and conformity. In that sense Stiegler's project can be seen as the very revival of critique called for by *The New Spirit of Capitalism*.

Notes

1. Hereafter I will refer to 'artist critique'. As Chiapello puts it, 'I prefer to speak of "artist critique" rather than "artistic critique" – especially since the latter is an ambiguous term liable to mean that artists are the subject of either the critique or its target' (2004: 586).
2. Of course, as Kate Nash among others has pointed out, there is a legitimate question to be asked here about how much an analysis of management literature can tell you about the actual attitudes of management cadres toward their work (Gilbert et al. 2010).
3. On the tendential understanding of capitalist crisis, see Noys 2010.
4. The use of the term *Eigentlichkeit* comes in the context of *Being and Time*'s discussion of *Dasein*'s 'ownmost possiblity'. As Macquarrie and Robinson, Heidegger's English translators, note, 'The connection between *eigentlich* ("authentic", "real") and *eigen* ("own") is lost in translation' (Heidegger 1962: 68 n. 3).

V: Pharmacology – The Poison that is also a Cure

15

Pharmacology and Critique after Deconstruction

Daniel Ross

One might be tempted to say [. . .] the thinking that was elaborated in France between 1960 and 2000 through these imposing personalities has left its heirs disarmed, and in a way inheriting it has undoubtedly led to a veritable sterilisation of thinking itself – often giving the impression of rationalising and legitimating the *renunciation of thinking an alternative,* by posing for example that there is in fact *no* alternative to the *state of affairs* that leads to universal unreason, no alternative other than 'resistance' to a kind of *inevitability of stupidity and performance,* imposed on us as a new regime in which knowledge has become an 'information commodity'.

Bernard Stiegler, *États de choc*[1]

Bernard Stiegler effectively poses to Jacques Derrida the following question: if deconstruction re-composes the 'binary oppositions' animating so-called metaphysics, what does that mean for the distinctions and differences we must continue to make? If Derrida drew attention to the Platonic diagnosis of writing as a *pharmakon* (the fact it contains both toxic and curative potential), and worried that the term 'deconstruction' itself contained the pharmacological threat of scholasticism, he nevertheless, perhaps, found it difficult to think how to counter this threat other than to call for *more* deconstruction. For Stiegler, the very great validity of the deconstructive style of thinking must not obscure that the future is always more important than the past, and thus that the question we must ask is what to *do* with and *make* of deconstruction, how to *invent* with it and what to *compose.*

Deconstruction contra critique

Derrida stated explicitly that 'deconstruction is neither an *analysis* nor a *critique*' (Derrida 1985: 3). Critique is on the contrary one object of deconstruction. Rodolphe Gasché reiterates that the view of deconstruction as a critical operation derives from insufficiently careful scrutiny of Derrida's text (Gasché 2007: 21). Deconstruction 'is an operation first and foremost upon the critical faith in the possibility of pure distinction and in the critical value of an immaculate, uncontaminated, invulnerable, and impenetrable limit'. On the other hand, neither is deconstruction 'anticritical':

> in spite of its interrogation of the possibilities of pure distinction, deconstruction recognises the necessity of distinguishing [. . .] But where such differentiation and discrimination takes place in the perspective of values of purity, fundamental originarity, and decidability, deconstruction questions the claims made. (2007: 22)

The point of difference between deconstruction and critique is the relation to 'purity'. Where critique *believes in* the possibility of purity, that is, in the possibility of 'mastering the threat of contamination', deconstruction, taking critique as its 'essential "theme" [. . .], is what Derrida calls "*différantielle* contamination", a thought that excludes the possibility of all rigorous distinction, of establishing itself in purity and without a contaminating relation to an other' (2007: 36). Gasché continues: 'This other is the referent against, from, and with which distinction occurs', an other that 'is infinite because it is irreducible', an other that is irreducible because it is not constituted through 'critical division' but rather that appears as what 'all critical decision presupposes' (2007: 36).

Deconstruction, responding to differance rather than instituting a decision, excludes the possibility of pure distinction because it strives to remain faithful to what precedes that distinction, which thus haunts distinction, that is, always threatens to contaminate any distinction. Whereas critique presupposes the pure autonomy of decision, deconstruction is generated by an originary difference and deferral, a process that Derrida insists is never merely negative. Now, while Bernard Stiegler accepts this analysis, he asks whether any critique has ever *truly* believed in the *purity* of distinction or the absolute exclusion of contamination. Can we really *oppose* deconstruction and critique? Stiegler contends that 'deconstruction failed to critique its critique of critique', to ask if critique is *necessarily* metaphysical, and failed, therefore, to ask 'what a critique might be *were it no longer founded on a system of oppositions*' (CPE, 15/26).

Deconstruction as composition

Stiegler inherits more from Derrida than from any other thinker: 'deconstructive' thinking is translated in Stieglerian terms into 'compositional' thinking:

> Deconstruction is a thinking of *composition* in the sense that composition is 'older' than opposition (what Simondon would have called 'a transductive relation': that is, a relation that constitutes its terms, the terms not existing outside the relation). It is a relation that is the vehicle of a process (that of *différance*), one very close, I would argue, to what Simondon also elaborates in terms of a 'process of individuation'. (DT, 249–50)

Deconstruction, pursuing the complex genesis of oppositional pairs, amounts to the elaboration of a process of becoming. It is therefore more consistent than first appearances might indicate with the theories of Gilbert Simondon, for whom the key was not to begin with terms or individuals and then to think their 'relation'; rather, it is the process itself that 'has the status of being' (Simondon 1992: 306). Stiegler's quotation marks around 'older', indicating that to some extent the reference is to a conceptual rather than temporal priority, should not obscure that Stiegler is describing the process through which distinctions arise, via the generative process of differance. Stiegler therefore argues that 'a thought *of individuation as process* is not foreign to that *process that differance also is*':

> 'To differ' is in this sense, which is that of differ*a*nce, to implement the *structural incompleteness* of the vital or psycho-social (but not mineral) individuation process such as it was thought by Simondon: 'this temporisation is also temporalisation and spacing, the becoming-time of space and the becoming-space of time'. (EC, 96 [§22], citing Derrida 1982: 8)

To this relation of differance to individuation should be added the influence of Nietzsche, for whom existence must be understood as a play of forces or, better, of tendencies. The formation of oppositions from prior compositions is an expression of this play of tendencies, as shown, precisely, by Nietzsche's critique of metaphysics:

> One of Nietzsche's most valuable contributions to the critique of metaphysics was his genealogy of guilt, insofar as guilt is a metaphysics that breaks with the tragic spirit by always and everywhere seeking the guilty, by *opposing* good and evil. It is necessary, in combat, for adversaries to *oppose* one another, but each of these adversaries is the representative, for their side, of a tendency that it cannot be a matter of eliminating, yet with which one must struggle. And adversaries represent, through their struggle, what Nietzsche himself called *eris*, or 'good discord', which is indeed an opposition, but which is also the way in which those tendencies represented by these adversaries compose a process, posing together, and one against the other, that process of which combat would be only a part, or rather, a *stage:* a stage of an individuation. Opposition, as the play of forces, plays out a more elementary composition, involving what Simondon called the *phase difference* inherent to the process of individuation. (DD1, 58/88)

And it may turn out that where compositional thought is superior to deconstructive thought is in making it possible to think *de*-composition, by showing that the decomposition of oppositions is not the same as that composition that is always already older than opposition, whereas deconstruction tends, perhaps, to see the deconstructive operation as the transformative restoration of originary complication. Deconstruction thus tends, perhaps, to perceive less clearly the pharmacological dangers of the destruction of 'oppositions', and the possibility that distinctions may on occasion be precisely what need to be preserved, that is, saved.

Différance and life

Stiegler accepts that all distinctions and oppositions emerge from a prior compositional complication, and it is from this perspective that we must understand the complication to which Stiegler draws attention in what he describes as his 'dialogue with Jacques Derrida around the concept of differance' (TT1, 136/147). Stiegler points to an unduly neglected aspect of the deconstructive project: when Derrida seeks to characterise grammatology as a positive science, not only does he draw upon the work of André Leroi-Gourhan in order to introduce a differantial account of the *grammē*, but he also states that differance is the history of life:

> Leroi-Gourhan no longer describes the unity of man and the human adventure thus by the simple possibility of the *graphie* in general; rather as a stage or an articulation in the history of life – of what I have called differance – as the history of the *grammē*. Instead of having recourse to the concepts that habitually serve to distinguish man from other living beings (instinct and intelligence, absence or presence of speech, of society, of economy, etc. etc.), the notion of *program* is invoked. It must of course be understood in the cybernetic sense, but cybernetics is itself intelligible only in terms of a history of the possibilities of the trace as the unity of a double movement of protention and retention. This movement goes far beyond the possibilities of the 'intentional consciousness'. (Derrida 1997: 84)

This reference to differance *as* the history of life relates, as Stiegler indicates, to the negentropic character of the biological tendency, 'a temporality that fights against disorder' (TDI, 155); life is a tendency running counter to that *overall* tendency we refer to as the 'law' of entropy. The history of life is the history of the genetic inscription of a difference and deferral of the entropic tendency of all matter. Difference and deferral: that is, a matter of the conservation of material forms, a form of 'memory' – the tendential persistence of genetic structure. Derrida, then, takes advantage of the generality of Leroi-Gourhan's concepts of exteriorisation and program in order to escape the metaphysics of humanism in his delineation of grammatology.

But is there not an inconsistency in Derrida's treatment of the relation between difference and life? *On the one hand*, Derrida refers to the 'most general concept of the *grammē*' and relates this to the entire negentropic rupture that is life, from the 'behavior of the amoeba or the annelid up to the passage beyond alphabetic writing [. . .], from the elementary programs of so-called "instinctive" behavior up to the constitution of electronic card-indexes and reading machines' (Derrida 1997: 84). Derrida thus describes a continuum in which hominisation would be one moment among others. *On the other hand*, he refers to difference as the difference and deferral of *physis*, as referring to 'all the others of *physis* – *tekhne, nomos, thesis*, society, freedom, history, mind, etc. – as *physis* different and deferred, or as *physis* differing and deferring' (Derrida 1982: 17). He seems in this case to be referring to a rupture in this history of life in which hominisation would announce the arrival of differance as the rupture *with* life.

Resolving this inconsistency depends for Stiegler on grasping that within the differantial, material, inscriptive process that is vital individuation, three great epochs of memory must be distinguished:

- that of genetic conservation, the persistence-in-becoming of the DNA molecule that has enabled the great terrestrial process of vital individuation;
- that of nervous memory, the capacity of animals possessing nervous systems to finitely retain, and to have their behavioural programs altered by, the events of their own experience;
- that of technical memory, the inscription of form in inanimate matter by beings whose cortical evolution is then affected by this capacity and the specular capacity to 'return' to these technical objects, which then also function as a projection screen and contribute to the formation of a non-biological process of becoming including the formation of socio-ethnic programs, idiomatic differences, technical inventions, and all the others of *physis*, the pursuit of life by means other than life (TT1, 17/31), amounting to what, following Simondon, Stiegler calls processes of psychic, collective and technical individuation.

Each of these epochs of memory (the genetic molecule, the neuronal structure, and the technical artefact) itself has a history of persistence and change, of 'metastability' (of *relative* stability, which also means, of relative *instability*). In particular, the third of these epochs, arising perhaps two million years ago, contains all the epochs of human technical history, including the distinction that occurred with the invention, around thirty thousand years ago, of artefacts that were not only technical but *mnemotechnical*, that is, deliberately and not accidentally recording instruments. This latter epoch is then subdivided by all those phase differences produced by the mnemo-technical changes wrought by everything from alphabetic writing to electronic card indexes, up until the process of total digitalisation unfurling today.

Derrida failed to conceive the necessity of thinking the singularity of each of these epochs, a grammatological failure that was also programmatological: if philosophy continues to have a task today, it must include the capacity to reflect not only on the past regimes and epochs of individuation, but on what vistas of individuation (that is, of differance) are opened up or closed off by the technical and mnemo-technical prostheses that ceaselessly proliferate in what is at present a technoscientific age of permanent innovation. This programmatological question is thus a question of *anticipation* in a world whose fundamental characteristic is that it changes more rapidly than our ability to comprehend it, where the speed of technical becoming exceeds that of culture (TT1, 15/29), and thus where the future *threatens* to be comprehensible only as an 'absolute danger [. . .], that which breaks absolutely with constituted normality' (Derrida 1997: 5).

Types of retention

Approaching such a question depends on being able to think the way in which the global retentional apparatus we have constructed also constructs us, that is, interacts with processes of psychic and collective individuation. Stiegler's critique of Derrida's hesitation in relation to differance must thus be composed with his critique of Derrida's deconstruction of Husserlian phenomenology. Husserl wished, without resorting to subjectivism or psychologism, to account for the 'idealities', for everything that does not seem to exist in the 'outside' world yet cannot simply be classed as belonging to the 'interior' world of the mind or consciousness. The 'existence' of the ideas, in so far as they can be neither dismissed nor reduced to elements of either the interior or the exterior, is in Husserl's terms 'transcendental' in the sense that ideas exceed any opposition between mental interiority and empirical exteriority. The question is: where *are* the ideas, but also, from where do they come, and how?

It was not until the 'Origin of Geometry' that Husserl's approach to this problem included a mnemo-technical aspect, in the form of writing and those technical instruments through which, quasi-mythologically but plausibly, geometry could arise. As Stiegler points out (DT, 241), if it was through an analysis of this argument that Derrida's work gets going, then nevertheless, for the very young Derrida, this resort by Husserl to a 'technicist explanation' (Derrida 2003: 167), to a 'purely technical genesis' (2003: 169) for the transcendental idealities, is indicative of a 'confused probabilism, and of a prephilosophical empiricism' (2003: 167). It was only by returning to this analysis in his 'Introduction' to Husserl's 'Origin', by re-tracing and re-inscribing his own response to Husserl's 'technicist explanation', that Derrida's thought was able to develop into what it eventually became: the differantial account of writing, that is, the deconstructive account of time. But for Stiegler, this account of temporality will prove to be fundamentally incomplete.

Husserl's greatest insight in *On the Phenomenology of the Consciousness of Internal Time* was that the problem of the perception of continual duration can only be resolved by recognising that the 'instant' of perception must contain within it the *retained* perception of the instant just past, 'as a comet's tail that attaches itself to the perception of the moment' (Husserl 1991: 37). The present 'instant' should thus be understood as a 'large now' (TT1, 246/251). The means by which Husserl achieved this insight was to focus not on the perception of time 'itself', but rather on that kind of object *in* the world that only exists *as* the passing from present to past. Husserl's exemplary 'temporal object' was the melody. It is only through the 'primary retention' that consists in the attachment of the sound 'just past' (1991: 38) to the present sound that the perception of a musical note or melody is possible. Husserl distinguishes between this primary retention and the secondary retention that consists in later recollection, as when, for example, we 'run through the melody in phantasy' (1991: 37).

Derrida's deconstruction of this 'absolute distinction' concentrates on the fact that however Husserl frames this retention or this attachment of the 'just past' to the present, this can only amount in some way to the inclusion of a non-perception *in* perception, of an absence in a presence. Husserl absolutely distinguished primary from secondary retention on the grounds that the former is somehow still present, a retention of what is both just past and still present, whereas the latter, which we understand as ordinary 'memory' or recollection, brings back to mind what is no longer present. Derrida argues that the past, however *just past* it may be, is by definition no longer present, and thus the difference between primary and secondary retention can only be a question of 'two modifications of nonperception [. . .], two ways of relating to the irreducible nonpresence of another now' (Derrida 1973: 65). His genius was to see that what this implies is not that between the present and the past is an unbridgeable gulf exceeding all analysis, but rather that it is the very idea of the present itself that must be deconstructed, that at the heart of the present must be an absence, and that the genuine distinction must be between re-tention and re-presentation. It is on this basis that Derrida was able to deconstruct the so-called metaphysics of presence by introducing the necessity of a constitutive and differantial absence at the heart of the living present.

That Husserl was trapped within the metaphysics of presence is further shown by his offhand dismissal of those traces constituted by forms of reproduction he called 'resembling objects', mnemo-technical 'depictions' such as paintings or busts (Husserl 1991: 61). But by the time Husserl comes to write 'The Origin of Geometry', this third kind of memory (which Stiegler will call 'tertiary') becomes the very means by which the fallibility and mortality of individual memory can be overcome, through which it turns out to have constitutive power. Derrida's observation that Husserl came to explain the genesis of idealities through the mnemo-technical articulation of the living with the dead eventually enabled him to think those 'quasi-transcendental' concepts of trace and archi-writing on which the grammatological project is founded.

Stiegler agrees that primary retention is inextricably intertwined with secondary retention, that is, that the perception I am having right now is not just a matter of being affected by the contents of what is perceptibly *there* before me, but is, on the contrary, a *selection* from those contents, the *criteria* for which derive from secondary retention, that is, accumulated prior experience. Perception is for this reason necessarily a *production*, a selective, editorial and even fictive *process*. Stiegler points out that for Husserl to develop his account of perception, the idea that it involves selection had to become possible, and that what probably caused it to occur to Husserl was the technical invention of the possibility of undergoing an identical perceptual stimulus *more than once*. This was the situation brought about by the invention of phonographic recording, which produces a kind of 'resembling object' in the form of recorded sound, and enables the listener to re-hear a reproduction of a temporal object such as a melody. The lesson of the phonogram is that an identical stimulus does *not* result in an identical perception, the difference between them being the result, above all, of having already listened to the recording and assimilated it to my stock of secondary retentions, that is, the result of the fact that I have changed.

But if Stiegler agrees that the *opposition* between primary and secondary retention must be deconstructed, for him the very same argument shows that what is occurring in any perception is the *interaction* of primary and secondary retention, that is, their *composition*. And this implies that, if they cannot be opposed, it is nevertheless vital that they be distinguished:

> The *difference* between primary and secondary retention is not a *radical* difference insofar as primary retention is unceasingly composed *with* secondary retention, that is to say, insofar as perception is always projected *by, upon,* and *in* imagination – contrary to what Husserl thinks, and Brentano as well. But it is no less the case that the *difference* remains and *constitutes* a distinction that is not an *op-position,* but precisely what I have called a *com-position.* (MS, 105)

And he continues:

> I am not saying that Derrida reduces the two Husserlian forms of retention to the same thing, but that he maintains in a certain sense that it is impossible to stipulate the difference between the two. And one cannot but be tempted to conclude that primary retention, in its difference from secondary retention, is in the end a vain distinction. I believe, on the contrary, that this difference is constantly playing itself out, as *differance, in that which tertiary retention projects and supports,* and as the magic skin of a will to power that is no longer that of a consciousness or a simple subject, but that is also a power and a knowledge – which compose while destroying themselves, by mutating, for instance, into technoscience. (MS, 105–6)

Intending and desiring

Stiegler, then, composes Husserl's account of the place of retention in temporal perception with his account of the place of tertiary retention in the origin of geometry. From this composition he draws the lesson that all perception and knowledge arises, for the individual as for the collective, from technically 'over-determined' processes of transmission and mediation, that is, that secondary and therefore primary retention always occurs within the conditioning context of the current mnemo-technical epoch. Just as technical prostheses made possible the origin of geometry, so too alphabetic writing and the book made possible the recapitulation of this origin in the minds of following generations of students of geometry. This recapitulation, which is a matter of the metastable re-constitution of idealities, is a very complex composition of primary and secondary retention in what we call the educational process, over-determined by tertiary retention. The result, for geometry as for every branch of knowledge, is – in so far as education *is* a process, and a process of (always incomplete) individuation, and an individuation process that consists in the continual *return* of the past (and the dead) in the (minds of living individuals in the) present – the constitution of what Stiegler calls a 'very large now' (DT, 246; TT2, 217/250).

But the epochality of this process does not cease with alphabetic writing or the printing press, extending, as Derrida says, to electronic card indexes and reading machines, and beyond. Now, while Derrida did indeed draw attention to the 'pharmacological' character of the mnemo-technical artefact through his reading of the Platonic diagnosis of the *pharmakon* of writing, his failure to pursue this history of supplements through its subsequent epochs meant that his work failed to provide instruments with which to think and critique the pharmacological character of those prostheses responsible for the conditioning of perception in the twentieth century. This industrialised perceptual conditioning, systematically devised by a capitalism striving to overcome its limitations by conquering markets no longer just territorially but 'spiritually', operates by systematically exposing minds to manufactured temporal objects such as television commercials, with the goal of significantly contributing to the secondary retentions of their audiences, that is, to their stock of memories. As these secondary retentions form the selection criteria of primary retention, this is a matter of conditioning perception itself.

But this is, of course, not just a question of 'perception' but of behaviour, specifically consumer behaviour. It is, therefore, a question of *desire*, of primary and secondary *pro*-tention, far beyond what is indicated by 'intentional consciousness'. In other words, the Husserlian account of perceptual selection must be corrected through an understanding of the technical over-determination of the process, but it must also be corrected through an understanding of the Freudian account of the effect of the perceptual process on that 'unconscious' that Freud referred to as the seat of desire.

At the same time, however, it must be noted that Freud himself did not take account of tertiary retention, paying little attention to the significance of radio and cinema in the constitution of contemporary desire (DD1, 106–7/147). Furthermore, Freud lacked an understanding of the temporality of the perceptual present, that is, he lacked the Husserlian distinction between primary and secondary retention, with the result that his attempt at a processual account of the way in which perception constitutes desire was unable to avoid getting stuck in the spatial metaphors that confuse Freud's account of trauma. Stiegler attempts to correct this with his description of the constitution of 'stereotypes' and 'traumatypes', which he uses to argue that psychic trauma can only affect the individual if it is in some way expected. The perceptual process involves anticipation and anticipatory *imagery* (DSL; see also Ross 2007: 204).

> The traumatism of the exterior is but the basis for the projection of a traumatype conserved in the interior but embedded in it and prevented from becoming conscious by the stereotypes, except when a pre-textuality causing primary retentional processes allows for the sudden freeing of the process of projection. Freud cannot see this: he is unable (just as Kant is unable) to distinguish between primary retentions and secondary ones. He must therefore oppose the inside and the outside. (DSL)

Stiegler thus attempts to think the 'selection process' operating in the interaction of primary and secondary retention through the prism of psychoanalysis as well as phenomenology, that is, to think the relation of intentionality and desire as forms of protention, as anticipation and motive.

If Derrida more than anyone helps to make possible this mutual deconstruction of Freud and Husserl, such that retention, intention and protention are informed by desire, motive and the unconscious, and vice versa, nevertheless Derrida does not himself think the role of tertiary retention in the constitution of desire, an absence that leaves his thought incapable of grasping the fundamental processes of contemporary capitalism. On the contrary, technics, the trace and so forth all tend in Derrida's work less to form part of a process of the constitution of desire than to be responsible for a kind of automaticity connected to the repetition compulsion. As such, technics tends to be assimilated with the death drive, with the threat to every principle, rather than grasped as constitutive of the process of principle-formation itself (Derrida 1995b: 12). It is to counter this disconnection of desire and technics that Stiegler is brought to write that 'the time has come to think technics on the basis of desire, and desire from out of technics' (MS, 109).

Invention and performativity

This difference plays out in Derrida's account of invention, which he refers to as that which, fundamentally, interrupts every process, that excess over and above every process that brings forth the absolutely new. Invention is thus the *origin* of the process. 'An invention always presupposes some illegality, the breaking of an implicit contract; it inserts a disorder into the peaceful ordering of things, it disregards the proprieties' (Derrida 2007: 1).

And yet Derrida also recognises that today invention is technoscientifically 'programmed', systematically and industrially planned and organised, that research and knowledge are nowadays elements of a 'programmatics of inventions' (2007: 27). Stiegler questions whether an understanding of invention as 'illegality' and 'disordering' is sufficient for thinking the technoscientific ordering of the inventive process on a global scale. He contrasts Derrida's description with Simondon's understanding of invention as an aspect of the system itself, and as founded on the protentive and anticipatory aspects of all perception: 'The retentions that form memory are constituted from images, from which protentions are projected, that is, anticipations, which are also images: time is made of images of the present, past and future' (EC, 135 [§31]). Invention is an operation founded on the fact that the editorial process underlying all perception is *necessarily* a generative, projective production of 'images', that is, motivated idealities:

> From such a perspective, *invention would be a reorganisation of the dynamic field and thus of the potential formed by the generator of mental images*, linked with object-images, on the ground of *a priori* images, occurring during a *cycle* that results in *new capacities for anticipation*. (EC, 137 [§31])

Invention is thus a form of anticipation generated by a tension in the metastable relation between the elements of a technical system or its relation to its milieu. The anticipatory mind, generator of images, manages to resolve such tensions by 'realising' its images in the form of technical invention. Such a perspective sees invention as ensuring and restoring the legality and order of the process, even though it is also what institutes phase differences *in* the individuation process.

Derrida's neglect of this aspect of invention, whereby it contributes to both the synchronic and diachronic moments of a system, is visible, perhaps, in his take on Austinian performativity. In 'Declarations of Independence', for example, with its brilliant account of the impossible way in which Jefferson's signature brings into being that which it presupposes ('the signature invents the signer' [Derrida 2002; 49]), Derrida again demonstrates the irruptive disorder underlying the performative inventiveness of political foundation. Stiegler himself refers to Derrida's treatment of performativity, but ties this more strongly to the performativity of knowledge elaborated by Jean-François Lyotard (DD1, 153/201; Lyotard 1984). The performativity of utterance is thus seen as part of the larger processes that are the anticipatory aspects of any technical system, but especially the contemporary technoscientific system that is premised not on the classical scientific notion of a 'description of the real' but rather on the performative 'inscription of a possible' (Stiegler 2007a: 41).

But even more than that, it is the *pharmacological* character of performativity to which Stiegler draws attention. On the one hand, performativity is a serious threat to the very existence of politics, 'synchronising' political understanding through a mediatised political spectacle using every technique of marketing to create 'audiences' and thereby destroy the *demos*. On the other hand, if politics is the attempt to find solutions to the problem of living together, then the invention of such solutions will always have a performative aspect, that is, be a matter of creating fictions. For one of the founders of Greek democracy, Cleisthenes, the solution to political disorder was the performative reinvention of socio-ethnic programs designed to increase Athenians' sense of belonging. The lesson to be learned from him is that democracy is always an *adoptive* process, that is, the creation of a fiction that is then *taken up* on the grounds it is *good* (DD1, 6–7/22–3). As Stiegler states, the problem today is to know how to distinguish good fictions from bad ones, precisely because the generalised performativity that is the consumerist adoptive process is more destructive than constitutive of desire (DD1, 148–9/195–6).

Critique after deconstruction

Finally, this fundamentally pharmacological character of invention and performativity must rebound upon deconstruction itself. If the lesson from Cleisthenes concerned not just the institution of a (technical) solution to a political impasse, but the *taking up* of that solution, that is, its (cultural) adoption, *belief* in that fiction as *making possible* what had hitherto been impossible, then this is to admit the necessity of the question of *will*. This is not a matter of some 'spirituality' opposing technics, but of what Stiegler refers to as the 'second redoubling' that takes up technics as one takes up a musical instrument, that is, in order not only to play but to practice it, and to compose with it, '*composing with the process,* [...] putting it to work *through my capacity for invention*' (AO, 73/74).

For Stiegler, there was a price to be paid for Derrida having conceived deconstruction less as a *decision* of thought than as a moment of a process of which it is simply an element, in some way 'automatic'. For this reason Stiegler, in *What Makes Life Worth Living*, asks what it is that deconstruction *wants*. And he answers: does it not, however much it insists on the inescapability of so-called metaphysics, want to escape the lures of 'metaphysics'? How could deconstructors not *want*, at some level at least, to *take up and take further* the Enlightenment project of conquering intellectual maturity, that is, autonomy, for individuals and collectives, but to do so, precisely, by not falling into the traps of transcendental subjectivity?

> But does such a claim not amount to hyper-criticism? To ask this question is to enter into and to claim to have undertaken a hyper-critique of the limits of deconstruction [...] Is it possible to reduce the pharmaco-logy of the *pharmaka*? Obviously not. Nobody has said better than Derrida why this is so. It is necessary to *make do*, or to *make the most of things* – that is, to make do with or make the most of the fact that *life is in the end ONLY worth living pharmacologically* [...] For in any case the deconstructor – who regularly claims the gesture of the *Aufklärer* in spite of everything that Derrida asserts *à propos* criticism and critique – would be unable to reduce the pharmacological condition that he deconstructs, which means that he himself projects lures, casts delusions, that he is unable to see. (WML, 76–7fr [§20])

If there remains something true about the idea that deconstruction was not a decision of thought, this may be because there is something about contemporary capitalism that amounts to a factual, material deconstruction (DT, 238). Stiegler's aim, then, is not to give up on deconstruction, but to take it up again, re-fashioned and re-armed, so that it is concerned less with the aporetic impossibility of every possibility, with simply 'resisting' that 'stupidity' that always wins, that is always on the side of the victor (Derrida 2009: 183), and more with constituting a critical project that would also be a *process and practice of invention*, on the grounds that *only* such a critique can undo the crisis that makes it possible (TT3, 152/227), but can do so, precisely, only in so far as it is taken up, that is, in so far as we adopt it, in so far as we succeed in believing in it. This is Stiegler's challenge.

Note

1. EC, 133 (§30)

Techno-pharmaco-genealogy

Stephen Barker

The Simondonian way of thinking *overcomes* the oppositions between types of individuation on the basis of traits common to all types of individuation processes [...] as *affiliations* and *relations* in what Simondon calls an ontogenesis, and which I prefer to call a genealogy.

Bernard Stiegler, *États de choc*[1]

Apeiro-technology

When Bernard Stiegler asserts, at the opening of *Technics and Time, 1*, that 'at its very origin and up until now, philosophy has repressed technics as an object of thought. Technics is the unthought' (TT1, ix), he is making both the immediate point that from Plato onwards technics has been the dangerous other, that only with Nietzsche, Freud, and their descendants has technics emerged from the shadows; and the subtler but more significant point that technics is both hyper-anthropological and extra-phenomenological. This grounding assumption – that technics *is* the unthought – provides the substructure for a vital aspect of the trajectory of Stiegler's work. This trajectory has consisted in a certain kind of genealogy, a genealogy of impact rather than of inheritance, that itself echoes the very nature of *tekhnē* as *savoir-faire* and *savoir-vivre* and only secondarily as the machinic. Stiegler's own thought and work, though it has been consistently oriented to *tekhnē*,[2] is the trace of a remarkable genealogical progression with regard to a constellation of nodes, including but hardly limited to *tekhnē*, from an early concentration on Derrida and Husserl in *Technics and Time, 1*, through the significant 'arrival' of Gilbert Simondon's work on (trans)individuation, to most recently a shift in emphasis from the work of Derrida to that of Deleuze, in, for example, *États de choc (States of Shock,*

2012). Given that all technological objects are pharmacological, a consistent guidepost for Stiegler, increasingly forceful in recent work, has been the axiological Platonic/Derridean focus on pharmacology and 'its' complex role *vis-à-vis* the question of technics. Stiegler will (like Heraclitus) treat the *pharmarkon* as *ambiguous* rather than endorse Plato's sense of it as *ambivalent*, in a distinction that emphasises the influences of the Deleuzian rhizomatic and Simondonian transduction. Addressing this genealogy, as clusters of force or impact, is essential to an understanding of Stiegler's development and his network of associations, the hyperdissemination of contemporary technics. My focus here will be on the nuanced way in which Stiegler relates Gilbert Simondon's treatment of (trans)individuation, transduction and collective individuation to technics in general and to the hypertechnics of contemporary culture; and how that relationship inevitably results in a post-Derridean conception of the pharmacological. The relationship Stiegler's work lays out for Simondon's key concepts is a function of a kind of (Heraclitean) cross-fertilisation: their interrelationship has a fundamental effect on all other nodes in the apparent structure.

The genealogy of the complex relational constellation resulting in Stiegler's concept of technicity revolves around a particular lodestar: the work of Gilbert Simondon and the two parts (published in three volumes) of his Sorbonne doctoral work, comprising the main thesis, *L'Individuation à la lumière des notions de forme et d'information* (*Individuation in Light of the Notions of Form and Information*), and the complementary thesis, *Du mode d'existence des objets techniques* (*On the Mode of Existence of Technical Objects*). While the latter was published immediately following his thesis defence in 1958 and reprinted in 1989, the former was eventually published in two volumes, the first in 1964 as *L'Individu et sa génèse physico-biologique* (*Individuation and its Physico-Biological Genesis*); the second, only twenty-five years later, as *L'Individuation psychique et collective* (*Psychic and Collective Individuation*, 1989).[3] The first volume of *Individuation* has only recently been reprinted, in 2012, while the second was reissued in 2007, at the instigation of Stiegler, who also wrote its preface.[4] This stalled history of publication gives some sense of the broader neglect of Simondon's work. Prior to Stiegler, his influence was felt mostly through Deleuze, who cites *Psychic and Collective Individuation* in *The Logic of Sense* (1969), in the

context of the relationship of the 'pre-individual' to the event
and of the 'limit of the living' – the 'polarised membrane' of
'in-side' and exteriority (Deleuze 2004b: 124 n. 3/126 n. 3) – and
more extensively in the slightly earlier *Difference and Repetition*
(1968), as he explores individuation as 'intensity' (Deleuze 2004a:
307/317).

Stiegler makes systematic yet subtle use of Simondon, capitalis-
ing on the development of Simondon's thinking and his thesis, creating a web of cross-fertilisations between 'the mode of existence
of technical objects', the explication of both psychic and collective
individuation, and his (Stiegler's) own eclectic thinking regarding
the *application* of Simondon's important work to twenty-first-
century philosophy and theory. Stiegler 'updates' Simondon just as
he does Derrida, using and going beyond the work of both.

Stiegler's conception of 'technological being' has its origins in
On the Mode of Existence of Technical Objects, where Simondon
states:

> The genesis of the technical object is part of its being. The technical
> object is what is not anterior to its becoming but is present at each
> stage of this becoming; the technical object is a unity of becoming
> [*unité de devenir*] [. . .] according to a principle of internal resonance.
> (1989: 20)

This description of the technical object as a 'unity of becoming'
refers to the way that the tool will evolve in relation, or 'adapt', to
its user's environment – either by finding new applications for its
use, or by being honed the better to fit the task for which it was
designed (1989: 50–5). For Simondon, as it will be for Stiegler,
the technical object is thus in a perpetual process of regeneration,
resulting from the impetus of its initial design, a reverberation or
inner energy as common to crystals as to human beings. Simondon
recognises that, in the case of living beings, this is not simply a
question of adaptation: 'the living being solves problems not only
by adapting, which is to say, by changing its relation to its milieu
[. . .] but by modifying itself, by inventing new internal structures'
(Simondon 2007: 17). Stiegler goes further still in rejecting the
language of adaptation, a term loaded with neo-Darwinist overtones of passively living or dying according to environmental
fit. He argues that technical evolution is a process of *adoption*,
meaning the way that tools are continually reinvented through

experience. Nonetheless, Stiegler's project develops directly out of the language Simondon uses to lay out his sense of technicity:

> The different aspects of the individualisation of technical being constitute the centre of an evolution that proceeds by successive stages, but which is not dialectical in the proper sense of the term, since the role of negativity is not its developmental engine. In the technical world, negativity is a fault in [*un défaut d'*] individuation, an incomplete conjunction of the natural and the technical world. (1989: 70)

The evolution of a tool, or technical object, is not determined by negation, in that it is not reactionary in any sense and does not operate in terms of a binary opposition of success (survival) and failure (death). It is rather a process of accumulation, through which new uses are revealed, and modifications made, in response to its environment. It thus 'passes through' diachronic stages within a synchronic framework. Simondon's designation for this developmental impetus is 'invention' (1989: 73) which, Simondon and Stiegler say, is a process of creation, or individuation, within an 'associated milieu', meaning an environment composed of interrelating technical objects. It is not simply tools that are individuated within these environments, but those who use them: just as the tool evolves in relation to the availability of other tools, the individual changes in response to the tool, discovers new modes of existence that were not previously available.

The set of conditions for technical evolution is based upon the energy of becoming within a metastable or 'associated' milieu, and operates within the context of a 'margin of indetermination' or unpredictability that prevents the individual from becoming an automaton. A technical being 'so perfect' (1989: 139) that its margin of indetermination – its capacity for self-transformation through the adoption of technical objects – is zero would manifest *bêtise*, being capable only of infinite repetition without change or development. The margin of indetermination in any given technical object or being consists of 'a number of critical phases' (1989: 141) within which the endless reception of information can be 'digested' and used toward transformation.[5] Given that this reception is a perpetual one, the origin or genesis of technicity (*through* technicity) is also perpetual, in accordance with what Simondon calls the '*dédoublement de la technicité* [. . .] *en figure et fond*', 'the doubled technicity of figure and ground', and what Stiegler calls 'doubled re-doubling', through which a consistent milieu allows for the potentially infinite application of technical gestures. This amounts to nothing less than the genealogy of technics.

The continuous process in which technical objects work 'through' human beings, resulting in the technical beings that we are, is the transductive ground and possibility – 'the support and the symbol' (1989: 247) – of a trans-individual relationality in which 'pure information' can become significant *only* within the system of technical objects. This process, its operation and its objects can be understood by 'technical beings' only within this context, since such contextualised (and perpetually re-contextualised) informa- tion, functioning within the transductive process, relinquishes any 'pure' advent in favour of a rapport between forms which, relative to the 'subject' that the process constructs, are simultaneously extrinsic and intrinsic. The process operates both temporally and spatially, in that it constructs both the individual and the collec- tive, the subject and the social.

An important link between Stiegler's and Simondon's techni- cal thought occurs through the centrality of what for Simondon is *totality* or *unity*, but which Stiegler (through Simondon) terms *metastability*. Simondon evokes a complex set of cross-genre references to describe the *technical individual* whose 'modernity' develops within the context of the industrial machine; that is, focused on the shift of the role of 'tool-carrier' from the human being to the machine, and thus the shift, as Stiegler points out, from technical individual to 'either the servant (worker) or the assembler (engineer or organiser)' (SM1, 101). The warning Stiegler offers is framed by the shift from the machinic to the hyper-machinic: Simondon's critique operates within the former, Stiegler's the latter; in both cases, however, it also touches upon their understanding of the aesthetic. For Simondon and Stiegler alike, the aesthetic object is not, properly speaking, an object, but rather what Simondon calls a 'prolongation' of the natural and the human worlds (1989: 187). In this sense, the aesthetic (pseudo-) object acts as a kind of force-field within which transformation, transduction and (trans)individuation occur, acting 'horizontally' (1989: 199) across genres, modes, milieus. The vital implication here is that the aesthetic object does not direct nor even guide the energy of phase-shifts, but rather provides a radically open passage from one domain to another, preventing the narrowing of modes of thought through what Simondon calls its potential for 'perpetual deviation' from a centralising or limiting pathway, establishing rather a 'transductive continuity' (1989: 199) that slips across ontological and epistemological limits and boundaries. This technical tendency evolves, in Stiegler's thinking, into his sense of the *pharmakon* itself.

It is necessary to say a word about what Simondon calls '*la pensée esthétique*', aesthetic thought, relative to transduction within and across phases in the evolving of the technical being. Simondon calls the movement from one phase of becoming to another '*déphasage*', meaning phase-shift or dephasing. Aesthetic thought is essentially the 'shift' in 'phase-shift', providing the neutral point (1989: 160) of an individuation that is not (and, because it re-creates the individual in and through each phase, cannot be) purposive. The fact that Simondon links aesthetic thought to the symbolic (that is, imagination) through invention makes it all the more central to Stiegler in *Technics and Time* and in works such as the two volumes of *De la Misère symbolique*, in which Stiegler thinks through the radical change in individuation brought about by hyperindustrialisation. Hyperindustrialisation describes a moment in the history of capitalism where technology and its users no longer stand in a relationship of mutual trans-formation; where individuals are produced homogeneously (or 'grammatised') by industrial technologies that they are unable to appropriate for the purpose of transforming themselves. The result is that the phase-shifts of technical evolution become a 'succession of losses of individuation' (MS1, 104), or *disindividuation*, the short-circuiting of long-term attention and of the ability of tertiary memory supports to give rise to a *positive* collective individua-tion. Stiegler thus shifts Simondon's aesthetic thought into what he describes as 'the axiological foundation of the grammatisation process' (MS1, 117), referring to a concept borrowed (and slightly reworked) from Sylvain Auroux. 'Grammatisation', here, means the standardisation of production technologies and a correspond-ing expropriation of consumers' ability to individuate themselves, or differentiate themselves from one another (DD1, 38–40; see also Auroux 1994: 120–2).

It is in the second part of Simondon's thesis that this concept of individuation, so crucial to Stiegler's own development, is laid out more fully. *L'Individuation psychique et collective* lays out the sense of individuation that provides the focus for Stiegler's work on the complex and troubled relationship between technics and the living being (*le vivant*).[6] For Stiegler, as for Simondon, the individuation process is 'negentropic', the accruing of information in tertiary retentions that will then frame *and* perpetually re-invent the discussion of the One and the Many as *hypokeimenon proton*, the primal support of enculturation. For Stiegler, 'culture is the intergenerational transmission of attentional forms invented in the course of individual experience, which becomes collective because psychosocial memory is technically exteriorised and supported' (Stiegler 2012a: 4). This means that 'individual experience' is endlessly dynamic, continually redefined by the 'attentional forms' embedded in culture. According to Stiegler, Simondon 'reactivates' the fundamentally social Pre-Socratic question of 'disindividuation' through *stasis* (Stiegler 2007b: vii), which Stiegler sees as the destruction of dynamic attention, nowadays carried out by its capture in the current web of hypermediation. The key to what goes on in the 'theatre of individuation' – which is the key to Stiegler's view of participation in culture in general – lies in the rapport between the 'pre-individual' and the technical tools through which it comes into being (2007b: iii). Individuation occurs within a 'field' of dynamic interactions that transcends (*dépasse*) the specific dyad of interior/exterior. In this respect, transductive individuation is always trans-individuation: 'individuation as transindividuation producing the transindividual' (2007b: xiii). The milieu of trans-individuation is described as an 'obscure zone' that is 'incompatible with itself' (2007: 11–12). In other words, the metastable operation of psycho-collective individuation takes place within the context of the initially unstable, ex-centric becoming of the individual.

Simondon's explication of the 'peripheral' character of the phase-element – that 'it' has no 'interiority' – provides the requisite opening for Stiegler's insights regarding tertiary retention. A vital element here is the *non-causal* nature of the transduction occurring in the obscure zone of the phase-shift: *trans*-duction is not *de*-duction nor *in*-duction, not causal relationships, but 'horizontal' ones that are 'not a synthesis, a return to unity, but the phase-shift of being out of its pre-individual centre of potentialised incompatibility' (2007: 27). In this regard, Simondon's transduction (and thus trans-individuation) re-constructs the very idea of ontogenesis, meaning the origin of the being of the individual. Stiegler calls attention to this ontogenetic process in his preface to the 2007 publication of *L'Individuation psychique et collective*, pointing out from the start that the individual comes into being 'between' the *who* and the *what*, with the knowledge it acquires through the adoption of tools also placing it in relation to a mythical origin, 'the fiction of an absolute past' (Stiegler 2007b: i). The interstitial process of transindividuation is 'paradoxically an irreducible inadequation between the act of knowing and the result of this knowledge' (2007b: ix). The negentropic accretion of information provides the ground on which phase-shifts occur, since information is what it is only as a *correlation* between source and receiver on 'the metastable field of *experience*' (Simondon 2007: 60), the former perpetually altering the latter.

It is only in the second chapter of *L'Individuation psychique et collective* that Simondon introduces the concept that opens his idea of transduction to Stiegler's use and further development of it: *apeiron*, the undetermined or unspecified (2007: 111), which 'can never be actualised in the *hic et nunc*' and which produces the '*angoisse*', the distress or anxiety that necessarily inhabits transduction and contributes to the perpetual problematics of subject-formation. For Simondon, the 'subject' is an 'ensemble', formed (and perpetually deformed and reformed) by the 'individuated individual and the *apeiron* he carries with him' (2007: 199) in the guise of the various kinds of 'signification' that can only be actualised collectively (that is, through sign-systems, of which language is only the most 'expressive' but by no means the only one):

> If there were no significations to support language there would be no language; language does not create signification; it is only what conveys between subjects an information that, in order to become significant, must encounter the *apeiron* associated with the individuality defined in the subject; language is the instrument of expression, the conveyer of information, but not the creator of significations. Signification is a relationship of beings, not a pure expression; signification is relational, collective, transindividual. (2007: 200)

Thus expression and information are *results* of a 'prior' indeterminacy, an *apeiron* both in terms of the preindividual and tertiary retention; Simondon's way of asserting that the 'subject' is never fixed is to say that 'ontogenesis is anterior to logic and ontology' (2007: 223), that 'the theory of individuation must thus be considered as a theory of phases of being, of its becoming as essential'. The (de)construction of signification, through the non-logic of phase-shifting through which individuation and trans-individuation occur, though it is metastable as a process, is not and cannot correctly be seen as a succession proceeding in serial form, but rather (given that it is predicated on the *apeiron*) a 'perpetually renewed re-construction (*résolution*)' that 'proceeds through crises', which is the very nature of ontogenesis. The individuated being of the subject is never substantial, never substance, but 'being put in question, being as problematic, divided, reunited, sustained in this problematics that it imposes as becoming, just as it constructs becoming' (2007: 224). Individuation and transindividuation, immersed in the collective, are not the development of an individuated being but the perpetual process of becoming, and the becoming-process, of individuation through the *apeiron* of phase-shifts. The 'being' does not pass through phases that thus modify him or her; being *becomes* being through phase-shifts whose nature is to de-phase – to de-center, as it were – the centreless. For Simondon, individuation is a matter of radical discontinuity, of what he refers to as 'the existence of thresholds' (2007: 273) or, indeed, individuated existence *as* thresholds: *apeiron* as *seuil*.

Stiegler's genealogical reworking of Simondonian ontogenesis also plays out through the concepts of grammatology (Derrida) and grammatisation (Auroux). For Derrida, originary *différance* is a principle entirely independent of the experiential; for Stiegler, grammatisation only occurs through technical objects, as mnemotechnics: the experiential is a consequence of technics *and* of its concretisation in technical objects. Stiegler refers to this as the 'originary prostheticity' of the experiential (as consciousness). Because for Derrida all semiological systems are distinct from 'arche-writing', difference precedes the technical. Stiegler radicalises this notion by insisting on the *a priori* technicity of 'organised inorganic matter': technics, in the form(s) of technical objects, provides the conditions for all inscription; for Stiegler, transductive inscription *is* mnemotechnics. As is made clear in *Ecologies of Television*, this equivalency means that Stiegler is in fundamental disagreement with Derrida regarding the relationship of technics and the artifactuality of technical objects within time. While for Derrida *différance* maintains an integrity that is affected by technics only in the sense that the technical is a form of the reification of *différance*, for Stiegler technics is the support for any organic subjectivity, the 'technical synthesis' providing the irreducible empirical base for human agency (ET, 158–60). Thus Stiegler operates within a double genealogy, both pushing Derridean *différance* 'back' a step, to technics, and re-orienting his genealogical project within the Simondonian context of the *apeiron*, shifting the Derridean *aporia* laterally. But this in turn leads to another genealogical threshold, from Simondon to Stiegler, since Simondonian preindividual forces, though they result in individuation and transindividuation, are not what *Stiegler* means by the tertiary and by tertiary memory: he means the inorganic organism, not the interiorised forces indicated by Simondon. In this sense, Simondon's very useful concept of 'mechanology',[7] the transduction of the human and the technical, is pushed further by Stiegler, who perceives both the force and the radical neutrality inherent in technics as he conceives it, a neutrality only hinted at – then retreated from – in Simondon, and even in Derrida, whose artifactuality remains inherently phenomenological.

It is only when Stiegler shifts his focus from Derridean *différance* to Deleuzian difference that his thought emerges as techno-*pharmacological*. This trajectory must be understood in terms of the reading, re-reading and reversal of the Platonic *pharmakon*. As Leonard Lawlor points out, though Derrida claims a 'nearly total affinity' between his work and that of Deleuze (Lawlor 2003: 67), in fact their treatment of Platonic difference marks the shift from a theory of opposition (Derrida) to one of multiplicity/neutrality (Deleuze) on which Stiegler will predicate his techno-pharmacology. The dialectics of poison/cure, constructed on a model of resemblance, that Derrida derives from Plato models itself on being, as Derrida says, 'intolerant in relation to [the] passage between the two contrary senses of the same word' (2003: 69; Derrida calls this 'contamination'), a model that severely limits the notion of *différance* since, as in Plato, in Derrida difference is subordinated to the same. For Deleuze, on the other hand, as Eric Alliez points out, 'the ontological power of becoming succeeds the pharmacological virtue of writing as an alogical milieu of forces. To be is to [. . .] be written in the ambiguous system of writing older than the arrested opposites *signified* by Plato' (Alliez 2003: 93). Thus the *pharmakon* operates as what Alliez calls a 'parricidal writing of signs' that is finally undecidable, that has 'nothing to do with the dialectical process'. In terms of the genealogy we are exploring, Deleuze takes a radically different tack from Derrida with regard to the 'undecidable', positing the pharmacological as *intensity*, radical neutrality, constructed on a model of multiplicity and dissimilarity (see Barker 2009). This is the genealogical step through technics to techno-pharmacology that informs Stiegler's current work, the chief factor in that recent work's theoretical (and political) power. Deleuze's own use of Simondon goes as far back as a short piece originally published in the *Revue philosophique de la France et de l'étranger*, in 1966. 'On Gilbert Simondon' focuses on what Deleuze calls Simondon's 'profoundly original theory of individuation implying a whole philosophy' (Deleuze 2004c: 86/115). Deleuze focuses on Simondon's principle of individuation based on 'a metastable system' that 'implies a fundamental *difference*, like a state of dissymmetry' (2004c: 87/116). Deleuze's 'dissymmetry' is precisely Simondon's '*déphasage*', as is clear in Deleuze's conclusion:

> What Simondon elaborates here is a whole ontology, according to which Being is never One. As pre-individual, being is more than one – metastable, superimposed, simultaneous with itself. As individuated, it is still multiple, because it is 'multiphased', 'a phase of becoming that will lead to new processes'. (2004c: 89/118)

The 'new process' to which Deleuze refers is what he calls an 'ethical' one, 'running from the pre-individual to the trans-individual' (2004c: 89/118);[8] it is an ethics based on an 'internal resonance' (2004c: 88/117) as 'information' – an ethics grounded in technics, but a new sense of the ethical emerging from the implications of tertiary retention as (the) *pharmakon*. This dephasing in Stiegler's trajectory is already significantly underway by the time it is openly declared in *États de choc*, where, in a discussion of the relationship between the human and the animal, Stiegler makes the contentious and perhaps hyperbolic claim that 'Derrida tends to reduce Deleuze's reasoning to a classic opposition between man and animal [. . .] I cannot prevent myself', Stiegler declares, 'from thinking that here Derrida *fait la bête*, totally misinterpreting Deleuze's discourse: he profoundly misunderstands the origin of the discourse on individuation (and disindividuation) in the repetition of *Difference and Repetition*' (EC, 60 [§15]).[9] Whatever one may think of Stiegler's position, Deleuze is very clear about *his* genealogical roots in Simondon, asserting that

> the essential process of intensive quantities is individuation [. . .] All individuality is intensive [. . .] comprising and affirming in itself the

difference in intensities by which it is constituted. Gilbert Simondon has shown recently that individuation presupposes a prior metastable state – in other words, the existence of a 'disparateness' such as at least two orders of magnitude or two scales of heterogeneous reality between which potentials are distributed. (Deleuze 2004a: 307/317)

For the Deleuze of *Difference and Repetition*, as for the Deleuze and Guattari of *A Thousand Plateaus*, Simondon provides, in the metastability of the preindividual, the groundwork for the development of the forceful neutrality he and they call 'intensive'. The 'disparateness' to which Deleuze refers is, *beyond* dialectics, 'an entire energetic materiality in movement, carrying *singularities* or *haecceities* that are already like implicit forms that are topological, rather than geometrical, and that combine with processes of deformation, [. . .] a zone of medium and intermediary dimension' (Deleuze and Guattari 2004b: 451). Simondon's instruction is that, as a materialist construction (transduction is never independent of the technics through which it operates), individuation is a process of *béance*, not of polarity but of the in-between, intensities of becoming passing through different 'phase-states', or what Deleuze would call stages of 'actualisation' (2004a: 255–62/266–73). For Stiegler, following Deleuze, the reversal of Platonism requires a re-thinking of the *pharmakon* as intensity, within the milieu of technics.

Techno-pharmacology

For us, who tend in the twenty-first century to remain non-inhuman beings, the question of the *pharmakon* is no longer only an academic concern occupying philosophy professors – it obsesses everyone. (WML, 15–16fr [Introduction])

The Stieglerian pharmakon, going beyond dialectical opposition, amalgamates the sense of technics Stiegler has been developing in his writing since the early 1990s, incorporating Simondon's transduction and (trans)individuation with a pharmacology initiated in Plato/Derrida and reaching its current nature – its radical neutrality – through Deleuzian immanence, the deterritorialisation of any and all duality. For Deleuze, territorialisations that 'collect or gather forces, either at the heart of the territory, or in order to go outside it [. . .] bring on a movement of absolute deterritorialisation' in which they 'cease to be terrestrial, becoming cosmic' (Deleuze and Guattari 2004b: 360). This concept of deterritorialisation, for Stiegler, describes:

> disarrangement between the social system and the technical system, because you are in a process. What is a process? A dynamic system; [. . .] technics is always pharmacological. It is always a *pharmakon*, because it is always creating a disequilibrium in the society in which it is developed and by which it is developed. (Stiegler 2011b: 41)

This is perhaps the most radical aspect of Stieglerian technopharmacology, its *unheimlichkeit* or undecidability. Unlike the toxic economic system that governs which technologies are adopted by consumers, as *pharmaka*, technologies themselves are not 'merely' ambivalent but profoundly neutral – which is to say, indifferent to the ways in which they individuate their users. Technics, *as* pharmacological in its emergence as information, only reveals its 'valence' when acted upon. This 'acting-upon' is itself an energy-exchange, whose direction is itself undecidable, as Simondon theorises through transduction. The transference from information to hyper-information, from technology to hyper-technology, far beyond the adaptive capacity of human beings (and as it has increasingly outstripped the rate of its adoption by the slowly-evolving human mind – in Stieglerian terms, the adoption rate at which the 'available brain' succumbs to it), is a new hyper-development of the originary philosophical question of sophistics and its reliance on hypomnemata. The energy-shifts enacted by and through the *pharmakon* are undecidable in advance, not merely binary, and thus more problematic than any binary, as Derrida clearly implies through the very nature of 'dissemination'.

In Plato's *Phaedrus*, writing is at first a 'useful', beneficial drug (*pharmakon*), then a detrimental, poisonous deadener of the soul (*pharmakon*). Socrates understands the *pharmakon* as the potential for language/writing to offer a more responsible alternative to sophistry; this in turn must remind us of its fundamentally technical nature, its applicability within the framework of Derridean grammatology and Stieglerian grammatisation, and to Deleuzian deterritorialisation in which *pharmaka* include all organological technics, including extra-linguistic social interactions such as family, school, politics. The implication is that *pharmaka* are inherently complex, multivalent artifices (perpetually shifting intensities) constituted from what Stiegler calls 'games and gambles' (Stiegler 2007c) that, as technological *practices*, consist of and in evolving discourses inherently interrogating the very nature of our participation in the cultural workplace, as we do the work of creating and recreating culture.

Stiegler's current strategy for employing the technopharmakon emerges through the exploration of the 'transitional object' of Donald Winnicott, which, along with Derrida's *pharmakon* and Deleuze's deterritorialisation, joins Simondon's transduction and (trans)individuation as an avatar of genealogy. Stiegler excavates this connection in *Ce qui fait que la vie vaut la peine d'être vécue: De la pharmacologie* (*What Makes Life Worth [the Pain of] Living: On Pharmacology*, 2010), where he reiterates the function of the *pharmakon* as *hypomnematon*, or 'artificial memory' (WML, 13fr [Introduction]). It is in the course of this exploration that the Stieglerian *pharmakon* evolves from an 'early', Derridean (structurally dyadic) phase (though even here Stiegler refers to it as an 'indefinite dyad' subject to 'the play of tendencies', WML, 115, 122fr [§§37, 39]), which, in producing attentional effects, is revealed as the essence of the psychotechnical. Stiegler then develops a radical new connection between technics and the *pharmakon*: 'the technical system links artificial organs that become the *pharmaka* of the psychosomatic body and that re-link it to other bodies within a social system' (WML, 185fr [§64]). The techno-pharmacological has thus emerged, seductive, playing within, among, across cultural strata, across economies of writing, critique, individuation. If the *pharmakon* is indeed the playful or ludic, 'the going or leading astray', as Derrida claims (Derrida 1981: 71), then coming to terms with it is all the more necessary for an understanding of Stiegler's most recent work.

For Stiegler, this elusive *nexus*, 'pharmacology', which is only by extension and misdirection an 'ology', is the lower layer of all these strata. Derrida introduces the *pharmakon* by reminding us that it is 'caught in a chain of signification'; Stiegler enters into the chain in the wake of Derrida's claim to have said all he '*meant to say*' (1981: 65). Stiegler's case is that we must *attend to* the flight of Derrida's subtle claims (and to Plato's ambivalences) regarding the *pharmakon* if we are to survive the hypermediated psycho-power of contemporary culture. We must question the genealogical principles of attentional forms in order to get to the essential issue for our time, which resides in the question of the new forms of metadata and of the original processes of transindividuation they allow us to envisage.

This is a genealogy whose mechanology, driving the process of transduction/individuation, is nothing less than 'a physical, biological, mental, or social operation' (Simondon 2007: 24) by means of which an activity circulates from one quasi-location to another, deriving this circulation from a perpetual re-structuring of the system as it shifts from one locus to another. For Simondon this linkage proceeds from region to region as a principle and a template, all modifications extending themselves progressively along with the operational structuring. On a larger scale, all information-transfers, in any medium, are transductive. Thus for Simondon transduction (like individuation) is a process of mediation, *en abîme*, subverting and interrogating all dualities, all dialectics, without eradicating them. Simondon (like Kant) creates a synthetic theory of perception, though for Simondon it is process: 'the conditions of possibility of knowledge are in fact the causes of existence of the individuated being' (2007: 127).

The Stieglerian technopharmakon disturbs dialectics from within, disturbs the entirety of the philosophical process (Plato acknowledges this conundrum, the abyss between the undecidable and the undecided). For Derrida, the *pharmakon* is the ground of *différance* with regard to a 'textuality constituted by differences and by differences from differences, by nature absolutely heterogeneous and constantly com-posing with the forces that tend to annihilate it' (1981: 98). Stiegler extends Derrida's cross-associations further, arguing that our genealogical connection to the Platonic critique of the *pharmakon* is to be found beyond textuality, beyond grammatology, in grammatisation. All tertiary memory, indeed all exteriorisation, results not only in knowledge, culture, memory – but in a contemporary hyperconsumerism that endeavors to take possession of the processes of transindividuation by dictating the ways in which transduction occurs. The transductive formation of attention, both strengthened and threatened by the *pharmakon*, is in perpetual flux, fleetingly achievable only through its inscription in collective individuation. Stiegler declares that 'there is no single *pharmakon*', by which he means that 'being' itself is technopharmacological. By connecting the genealogical arc from Simondon to Stiegler, narrating a concept of the *pharmakon* in its developmental (de)phases, Stiegler works through an ever-shifting technopharmacological understanding of contemporary culture – indeed of the nature of what he calls the 'non-inhuman' itself.

Notes

1. EC, 95–6 (§22)
2. 'Disorientation', the subtitle of the second volume of *Technics and Time*, is a key determinant of Stiegler's thinking through of *tekhnē* and technics: 'location' is always relational, determinable only through the proximate and according to Heisenbergian principles linking relationality and indeterminacy.
3. At the time of writing, none of these have appeared 'officially' in English, though an initial (partial) translation of the first section of *On the Mode of Existence of Technical Objects* from 1980 can be found at <http://english.duke.edu/uploads/assets/Simondon_MEOT_part_1.pdf> (last accessed 4 March 2013). The translation is by Ninian Mellamphy. Citation of the preface, by John Hart, is indicated by Roman numerals. Another substantial translation, by Taylor Adkins, can be found on the web at <http://fractalontology. wordpress.com/2007/10/03/translation-simondon-and-the-physico- biological-genesis-of-the-individual/>.
4. Interestingly, Stiegler's preface to the volume, enigmatically entitled 'L'Inquiétante étrangeté de la pensée et la métaphysique de Pénélope' ('The Disquieting Strangeness of Thought and the Metaphysics of Penelope') though it does not mention Penelope, opens with a declaration that it was François Laruelle who 'made Simondon's work known to me' (Stiegler 2007b: i). Laruelle's work, particularly on 'non-philosophy', which Laruelle simply calls 'science' (or *'science (de) l'Un* [science of (the) One]'), attempts to step 'behind' the 'decision' on which, according to him, all philosophy is based, avoiding the reflection that throughout the history of philosophy it has relied and been built on dialectical division. In the sense that Laruelle champions a notion of 'science' that is anterior to decision, he is working at the purely theoretical level of Simondon's pre-individual. Laruelle's non-philosophy questions the survivability of humanity, exploring how it/we must operate through a 'radical immanence' that pushes back against the metaphysical dualities underlying all philosophy, opting instead for a science of 'disalienated' truth that is 'more real' than philosophy. Laruelle's engaged theory informs Stiegler's work through Ars Industrialis (<http://www.arsindustrialis.org>) and the École de philosophie d'Épineuil-le-Fleuriel (<http://pharmakon.fr/wordpress/>).

5. Simondon points out that this process distinguishes between what Bergson calls 'open' and 'closed' machines (1989: 141).

6. Simondon distinguishes the systematic preindividuality of crystals from that of living beings (*vivants*):

> the living being cannot be compared to an automaton that would maintain a certain number of balances or would search for numerous compatibilities among numerous requirements according to a complex formula of balance composed of simpler formulas of balance. (2007: 17)

In the living being metastability results not from formulas of balance but from the *a priori* nature of *déphasage*, phase-shift, predicated on the emergence of the dynamic phase which distinguishes the pre-individual from the initiation or activation of the individuation process.

7. Simondon's critique of the technical object must thus be seen as praise of the machine and its sharing of ontological space with 'the human'. Simondon here calls directly upon the work of Jacques Lafitte, whose *Réflexions sur la science des machines* (Paris: Bloud & Gay, 1932), is the first to place the technical object on the elevated plane Simondon finds even more relevant in a cybernetic age. Simondon's address to the second conference on mechanology claimed that 'there is something eternal in a technical schema [. . .] which is always present and which can be conserved in a thing' (Gilbert Simondon, in *Cahiers du Centre Cultural Canadien*, no. 4, 'Deuxième colloque sur la mecanologie' (1976: 87), cited in John Hart's Preface to the Mellamphy translation of *On the Mode of Existence of Technical Objects*, 1989: ii). Hart refers to Simondon's work as 'the second moment in the emergence of mechanology' (1989: ii), following Lafitte's invention of the phrase; Stiegler's adoption of Simondon's concept of transduction – but in a more radical form – is its genealogical heir.

8. Deleuze quotes and corroborates Simondon's assertion that 'ethics exists to the extent that there is information, in other words, signification overcoming a disparition of the elements of being, such that what is interior is also exterior' (2004c: 89/118), thus contributing a vital element to the Stieglerian techno-pharmakon.

9. I have left this in French (Stiegler italicises the phrase) since the many nuances of '*faire*' and '*bête*' are precisely what Stiegler explores in *États de choc*, with its play on *bête, bêtise, stupide*, and stupid, through the development of the idea that 'it is in the *pharmakon* that the resources for struggling against stupidity must be found' (EC, 60 [§15]).

Notes on Contributors

Stephen Barker is Associate Dean in the Claire Trevor School of the Arts at the University of California at Irvine. Focusing on critical, aesthetic and performance theory, he has written on Nietzsche, Derrida, Beckett, Blanchot, Lyotard, Artaud and others, and translated numerous works by Bernard Stiegler for Stanford. His books include *Autoaesthetics: Strategies of the Self After Nietzsche* (Humanities Press, 1992) and, as editor, *Signs of Change: Premodern→Modern→Postmodern* (SUNY, 1996) and *Excavations and Their Objects: Freud's Collection of Antiquity* (SUNY, 1996). Another book, *Thresholds: The Art of Limit-Play*, is forthcoming.

Richard Beardsworth is Professor in the Department of Politics and International Relations, Florida International University. He is the author of books including *Derrida and the Political* (Routledge, 1996), *Nietzsche* (Les Belles Lettres, 1997) and *Cosmopolitanism and International Relations Theory* (Polity, 2011). He is co-translator (with George Collins) of *Technics and Time, 1: The Fault of Epimetheus*, and has written several articles on Stiegler.

Miguel de Beistegui is Professor of Philosophy at Warwick University, and author of numerous monographs on twentieth-century philosophy, literature and art, including *Heidegger and the Political* (Routledge, 1998), *Truth and Genesis: Philosophy as Differential Ontology* (Indiana, 2004), *Immanence and Philosophy: Deleuze* (Edinburgh, 2010), *Proust as Philosopher: The Art of Metaphor* (Routledge, 2012) and *Aesthetics After Metaphysics: From Mimesis to Metaphor* (Routledge, 2012).

Patrick Crogan teaches film and media at the University of the West of England. He guest-edited the special issue of *Cultural Politics* on Stiegler (2010), and co-edited the 'Paying Attention' issue of *Culture Machine* (2012). Author of *Gameplay Mode: War, Simulation and Technoculture* (2011), he works across film, animation and digital media forms of what Stiegler calls the industrial temporal object.

Martin Crowley is Reader in Modern French Thought and Culture at the University of Cambridge. His most recent book is *L'Homme sans: Politiques de la finitude* (Lignes, 2009). Other publications include *The New Pornographies: Explicit Sex in Recent French Fiction and Film* (co-authored with Victoria Best; Manchester University Press, 2007); *Robert Antelme: L'Humanité irréductible* (Lignes, 2004); *Robert Antelme: Humanity, Community, Testimony* (Legenda, 2003), and *Duras, Writing, and the Ethical: Making the Broken Whole* (Oxford University Press, 2000).

Oliver Davis is Associate Professor of French Studies and Programme Director of the Centre for Research in Philosophy, Literature and the Arts at Warwick University. His publications include the critical introduction *Jacques Rancière* (Polity, 2010) and the edited volume of critical essays *Rancière Now* (Polity, forthcoming 2013).

Tania Espinoza recently obtained a PhD from the Department of French, University of Cambridge, on the figure of negative space in philosophy, psychoanalysis and modernist literature. She has taught in the Department of Literature of the Universidad Mayor de San Andres, La Paz, and is, in 2013, a guest lecturer at ECLA Bard, Berlin. She is a member of the Société Internationale de Philosophie et Psychanalyse.

Sophie Fuggle is Lecturer in French, Nottingham Trent University. Her book *Foucault/Paul: Subjects of Power* is forthcoming with Palgrave Macmillan.

Christina Howells is Professor of French at the University of Oxford and a Fellow of Wadham College. She works on twentieth-century French literature and thought, Continental philosophy and literary theory. Her publications include *Sartre: The*

Necessity of Freedom (Cambridge University Press, 1988); *The Cambridge Companion to Sartre* (Cambridge University Press, 1992); *Derrida: Deconstruction from Phenomenology to Ethics* (Polity 1998); *French Women Philosophers* (Routledge, 2004); and *Mortal Subjects: Passions of the Soul in Late Twentieth-Century French Thought* (Polity, 2011).

Ian James is a Fellow in Modern Languages at Downing College and a lecturer in the Department of French at the University of Cambridge. He is the author of *Pierre Klossowski: The Persistence of a Name* (Legenda, 2000), *The Fragmentary Demand: An Introduction to the Philosophy of Jean-Luc Nancy* (Stanford University Press, 2006), *Paul Virilio* (Routledge, 2007) and *The New French Philosophy* (Polity, 2012).

Christopher Johnson is Professor of French at the University of Nottingham and a specialist of post-war French thought. He is author of *System and Writing in the Philosophy of Jacques Derrida* (1993), *Claude Lévi-Strauss: the Formative Years* (2003), and a monograph on André Leroi-Gourhan (forthcoming). He is a founding member of the Nottingham Science Technology Culture Research Group.

Michael Lewis is Senior Lecturer in Philosophy at the University of the West of England. He is the author of *Heidegger and the Place of Ethics* (Continuum, 2005), *Heidegger Beyond Deconstruction* (Continuum, 2007), *Derrida and Lacan: Another Writing* (Edinburgh, 2008) and (with Tanja Staehler) *Phenomenology: An Introduction* (Continuum, 2010). He is working towards a book on the contemporary reinvention of philosophical anthropology.

Gerald Moore is Lecturer in French in the School of Modern Languages and Cultures, Durham University. He is the author of *Politics of the Gift: Exchanges in Poststructuralism* (Edinburgh, 2011), as well as articles on recent French thought, psychoanalysis and literature (Michel Houellebecq). He is currently preparing a monograph, *Bernard Stiegler: Philosophy in the Age of Technology*, for Polity.

Ben Roberts is Lecturer in Media Studies at the University of Bradford. He has recently edited a special issue of the journal *New*

Formations dedicated to the work of Bernard Stiegler, and is currently working on a book project called *Contemporary Critical Theory and Technology*.

Daniel Ross is co-director of the film *The Ister* (2004), author of *Violent Democracy* (Cambridge University Press, 2004) and translator of five books by Bernard Stiegler: *Acting Out* (Stanford University Press, 2009), *For a New Critique of Political Economy* (Polity Press, 2010), *The Decadence of Industrial Democracies* (Polity Press, 2011), *Uncontrollable Societies of Disaffected Individuals* (Polity Press, 2012) and *What Makes Life Worth Living: On Pharmacology* (Polity Press, 2013).

Serge Trottein is Chargé de recherche at the Centre National de la Recherche Scientifique (UPR 76, Villejuif, France) and Associate Director of THETA (Théories et Histoire de l'Esthétique, du Technique et des Arts). He is editor of *L'Esthétique naît-elle au XVIIIe siècle?* (PUF, 2000) and is currently preparing a book on the function of aesthetics in postmodern philosophy.

Bibliography

Adorno, Theodor W. (1973) [1964], *The Jargon of Authenticity*, trans. Knut Tarnowski and Frederic Will, Evanston: Northwestern University Press.

Adorno, Theodor W. and Max Horkheimer (1979) [1944], *Dialectic of Enlightenment*, trans. John Cumming, London: Verso.

Alliez, Eric (2003), 'Ontology and Logography: The Pharmacy, Plato, and the Simulacrum', trans. Robert Rose and Paul Patton, in *Between Deleuze and Derrida*, ed. Paul Patton and John Protevi, London: Continuum, pp. 84–97.

Artières, Philippe, with Laurent Quéro and Michelle Zancarini-Fournel, eds (2003), *Le Groupe d'Information sur les Prisons: archives d'une lutte, 1970–72*, Paris: IMEC.

Assoun, Paul-Laurent (1990), 'L'Arsenal freudien', in *Corps écrit*, 35: 51–62.

Auroux, Sylvain (1994), *La Révolution technologique de la grammatisation: Introduction à l'histoire des sciences du langage*, Liège: Margada.

Barker, Stephen (2009), 'Transformation as an Ontological Imperative: The [Human] Future According to Bernard Stiegler', *Transformations*, 17, <http://www.transformationsjournal.org/journal/issue_17/article_01.shtml> (last accessed 18 December 2012).

Barthes, Roland (1980), *La Chambre claire: note sur la photographie*, Paris: Gallimard.

Bazin, André (1967), *What is Cinema? Volume 1*, trans. Hugh Gray, Berkeley: University of California Press.

Beardsworth, Richard (1995), 'From a Genealogy of Matter to a Politics of Memory: Stiegler's Thinking of Technics', *Tekhnema: Journal of Philosophy and Technology*, 2: 85–115, <http://tekhnema.free.fr/contents2.html> (last accessed 21 March 2013).

—(1996a), *Derrida and the Political*, London: Routledge.

—(1996b), 'Nietzsche, Freud and the Complexity of the Human', *Tekhnema: Journal of Philosophy and Technology*, 3: 113–40.

—(1998a), 'Thinking Technicity', *Cultural Values*, 2.1: 70–86.

—(1998b), 'Towards a Critical Culture of the Image: J. Derrida and B. Stiegler's *Échographies de la télévision*', *Tekhnema: Journal of Philosophy and Technology*, 4: 114–40.

—(2010), 'Technology and Politics: A Response to Bernard Stiegler', *Cultural Politics*, 6.2: 181–99.

Beller, Jonathan (2006), *The Cinematic Mode of Production: Attention Economy and the Society of the Spectacle*, Lebanon, NH: University Press of New England.

Bennington, Geoffrey (1996), 'Emergencies', *Oxford Literary Review*, 18: 175–216.

Bollas, Christopher (1987), *The Shadow of the Object: Psychoanalysis of the Unthought Known*, New York: Columbia University Press.

Boltanski, Luc and Ève Chiapello (2005), *The New Spirit of Capitalism*, trans. Gregory Elliott (London: Verso); *Le Nouvel Esprit du capitalisme*, Paris: Gallimard, 1999.

Bowie, Malcolm (1987), *Freud, Proust and Lacan: Theory as Fiction*, Cambridge: Cambridge University Press.

Buchloh, Benjamin H. D. (2000), 'Beuys: The Twilight of the Idol, Preliminary Notes for a Critique (1980)', in *Neo-Avantgarde and Culture Industry: Essays on European and American Art from 1955 to 1975*, Cambridge, MA: The MIT Press, pp. 41–64.

Budgen, Sebastian (2000), 'A New "Spirit of Capitalism"', *New Left Review*, 1: 149–56, <http://newleftreview.org/II/1/sebastian-budgen-a-new-spirit-of-capitalism> (last accessed 21 March 2013).

Capps, Donald (2010), '"Your Nose Has Again Smelled Right": Reflections on the Disputed Origin of Freud's Concept of Sublimation', *American Imago*, 67.2: 263–92.

Chiapello, Ève (2004), 'Evolution and co-optation', *Third Text*, 18.6: 585–94.

Christiansen, Morten H. and Simon Kirby, eds (2003), *Language Evolution*, Oxford: Oxford University Press.

Colony, Tracy (2010), 'A Matter of Time: Stiegler on Heidegger and Being Technological', *The Journal of the British Society for Phenomenology* 41.2: 117–31.

—(2011), 'Epimetheus Bound: Stiegler on Derrida, Life, and the Technological Condition' in *Research in Phenomenology*, 41.1: 72–89.

Crogan, Patrick (2010), 'Bernard Stiegler: Philosophy, Technics, and Activism', *Cultural Politics*, 6.2: 207–28.

Crogan, Patrick and Samuel Kinsley (2012), 'Paying Attention: Towards a Critique of the Attention Economy', *Culture Machine*, 13: 1–29.

Dawkins, Richard (2006) [1976], *The Selfish Gene*, 30th anniversary edition, Oxford: Oxford University Press.

Deleuze, Gilles (1995) [1990], *Negotiations 1972–1990*, trans. Martin Joughin, New York: Columbia University Press; *Pourparlers*, Paris: Minuit, 2003.

—(1999), *Foucault*, trans. Séan Hand, London: Continuum; *Foucault*, Paris: Minuit, 1986.

—(2004a), *Difference and Repetition*, trans. Paul Patton, London: Continuum; *Différence et repetition*, Paris: PUF, 1968.

—(2004b), *The Logic of Sense*, trans. Mark Lester and Charles Stivale, London: Continuum; *Logique du sens*, Paris: Minuit, 1969.

—(2004c), 'On Gilbert Simondon', in *Desert Islands and Other Texts 1953–1974*, trans. Michael Taormina, ed. David Lapoujade, New York: Semiotext(e), pp. 86–9; 'Sur Gilbert Simondon', *Revue philosophique de la France et de l'étranger*, 156.1–3, janvier–mars 1966: 115–18.

—(2005) [1983], *Cinema 1: The Movement-Image*, trans. Hugh Tomlinson and Barbara Habberjam, London: Continuum.

—(2006), *Nietzsche and Philosophy*, trans. Hugh Tomlinson. New York: Columbia University Press; *Nietzsche et la philosophie*, Paris: PUF/Quadrige, 1962.

Deleuze, Gilles and Félix Guattari (2004a) *Anti-Oedipus*, trans. Robert Hurley, Mark Seem and Helen R. Lane, London: Continuum; *L'Anti-Œdipe: Capitalisme et schizophrénie, 1*, Paris: Minuit, 1971.

—(2004b), *A Thousand Plateaus*, trans. Brian Massumi, London: Continuum; *Mille plateaux: Capitalisme et schizophrénie, II*, Paris: Minuit, 1980.

Derrida, Jacques (1973) [1967], *Speech and Phenomena and Other Essays on Husserl's Theory of Signs*, trans. David B. Allison, Evanston: Northwestern University Press.

—(1981) [1972], 'Plato's Pharmacy', in *Dissemination*, trans. Barbara Johnson, Chicago: University of Chicago Press.

—(1982), 'Différance', in *Margins of Philosophy*, trans. Alan Bass, Chicago: Chicago University Press; *Marges – de la philosophie*, Paris: Minuit, 1972.

—(1985), 'Letter to a Japanese Friend', in *Derrida and Différance*, David Wood and Robert Bernasconi, eds, Evanston: Northwestern University Press.

—(1995a), 'Avances', preface to Serge Margel, *Le Tombeau du Dieu artisan*, Paris: Minuit.

—(1995b), *Archive Fever: A Freudian Impression*, trans. Eric Prenowitz, Chicago and London: University of Chicago Press.

—(1997), *Of Grammatology*, corrected edition, trans. Gayatri Chakravorty Spivak, Baltimore: Johns Hopkins University Press; *De la grammatologie*, Paris: Minuit, 1967.

—(2002) [1976], 'Declarations of Independence', in *Negotiations: Interventions and Interviews, 1971–2001*, trans. Elizabeth Rottenberg, Stanford: Stanford University Press.

—(2003) [1954], *The Problem of Genesis in Husserl's Philosophy*, trans. Marian Hobson, Chicago and London: University of Chicago Press.

—(2006), *On Touching – Jean-Luc Nancy*, trans. Christine Irizarry, Stanford: Stanford University Press; *Le Toucher, Jean-Luc Nancy*, Paris: Galilée, 2000.

—(2007) *Psyché. Inventions of the Other, II*, ed. and trans. Peggy Kamuf and Elizabeth Rottenberg, Stanford: Stanford University Press; *Psyché. Inventions de l'autre, II*, Paris: Galilée, 2003.

—(2008) [2006], *The Animal That Therefore I Am*, ed. Marie-Louise Mallet, trans. David Wills, New York: Fordham University Press.

—(2009) [2001–2], *The Beast and the Sovereign, Volume I*, trans. Geoffrey Bennington, Chicago: University of Chicago Press.

Derrida, Jacques and Elisabeth Roudinesco (2004) [2001], *For What Tomorrow ... A Dialogue*, trans. Jeff Fort, Stanford: Stanford University Press.

Doane, Mary Ann (2002), *The Emergence of Cinematic Time*, Cambridge, MA: Harvard University Press.

Dubos, Jean-Baptiste (1748) [1719–33], *Critical Reflections on Poetry, Painting and Music*, trans. Thomas Nugent, 2 vols, London: John Nourse; *Réflexions critiques sur la poésie et sur la peinture*, ed. Dominique Désirat, Paris: École nationale supérieure des beaux-arts, 1993.

de Duve, Thierry (1988), 'Joseph Beuys, or The Last of the Proletarians', *October*, 45: 47–62.

Eagleton, Terry (1991), *The Ideology of the Aesthetic*, Oxford: Blackwell.

Edelman, Lee (2004), *No Future: Queer Theory and the Death Drive*, Durham, NC: Duke University Press.

Foucault, Michel (1972) [1969], *The Archaeology of Knowledge*, London: Tavistock.

—(1977) [1971], 'Nietzsche, Genealogy, History', in *Language, Counter-Memory, Practice: Selected Essays and Interviews*, ed. Donald F. Bouchard, Ithaca: Cornell University Press, pp. 139–64.

—(1980) [1977], 'The Eye of Power', in *Power/Knowledge: Selected*

Interviews and Other Writings 1972–1977, ed. Colin Gordon, New York: Pantheon, pp. 146–65.

—(1990a) [1976], *The History of Sexuality, 1: An Introduction*, trans. Robert Hurley, New York: Vintage Books.

—(1990b) [1984], *The History of Sexuality, 2: The Use of Pleasure*, trans. Robert Hurley, New York: Vintage Books.

—(1990c) [1984], *The History of Sexuality, 3: The Care of the Self*, trans. Robert Hurley, New York: Vintage Books.

—(1995) [1975], *Discipline and Punish: The Birth of the Prison*, trans. Alan Sheridan, New York: Vintage Books.

—(1997) [1983], 'Self-Writing', in *Essential Works of Foucault, 1: Ethics, Subjectivity and Truth*, ed. Paul Rabinow, New York: New Press, pp. 207–22.

—(2001a), *The Order of Things*, London: Routledge; *Les Mots et les choses*, Paris: Gallimard, 1966.

—(2001b), *Dits et Écrits 2 (1976–1988)*, ed. Daniel Defert and François Ewald, Paris: Quarto/Gallimard.

—(2005), *The Hermeneutics of the Subject: Lectures at the Collège de France 1981–1982*, trans. Graham Burchell, New York: Palgrave Macmillan.

—(2006), *Psychiatric Power: Lectures at the Collège de France, 1973–1974*, trans. Graham Burchell, New York: Palgrave Macmillan.

—(2008), *Le Gouvernement de soi et des autres. Cours au Collège de France 1982–1983*, Paris: Seuil.

—(2009), *Security, Territory, Population: Lectures at the Collège de France 1977–1978*, trans. Graham Burchell, New York: Palgrave Macmillan, 2009.

Freud, Sigmund (1955a) [1895], 'The Psychotherapy of Hysteria', in *The Standard Edition of the Complete Psychological Works of Sigmund Freud, Volume II (1893–1895)*, ed. and trans. James Strachey, London: The Hogarth Press, pp. 253–305.

—(1955b) [1920], 'Beyond the Pleasure Principle', in *SE XVIII (1920–1922)*, ed. and trans. James Strachey, London: The Hogarth Press, pp. 7–64.

—(1957) [1914] 'On Narcissism: An Introduction', in *SE XIV (1914–1916)*, ed. and trans. James Strachey, London: The Hogarth Press, pp. 30–59.

—(1958) [1912], 'Recommendations to Physicians Practising Psycho-Analysis', in *SE XII (1911–1913)*, ed. and trans. James Strachey, London: The Hogarth Press, pp. 109–20.

—(1961) [1930], 'Civilization and Its Discontents', in *SE XXI,* ed. and trans. James Strachey, London: The Hogarth Press, pp. 57–145.

—(1962) [1893], 'Charcot', in *SE III (1893–1899)*, ed. and trans. James Strachey, London: The Hogarth Press, pp. 7–23.

Gasché, Rodolphe (2007), *The Honor of Thinking: Critique, Theory, Philosophy*, Stanford: Stanford University Press.

Geller, Jay (2009), ' "Of Snips... and Puppy Dog Tails": Freud's Sublimation of *Judentum*', *American Imago*, 66.2: 169–84.

Gilbert, Jeremy et al. (2010), 'The New Spirit of Capitalism', *Soundings*, 45, <http://www.questia.com/library/1G1-237137543/the-new-spirit-of-capitalism-roundtable-discussion> (last accessed 21 March 2013).

Hadot, Pierre (1989), 'Réflexions sur la notion de "culture de soi" ', in *Michel Foucault philosophe*, Paris: Seuil.

Haig, David (2007), 'Weismann Rules! OK? Epigenetics and the Lamarckian Temptation', *Biology and Philosophy*, 22.4: 415–28.

Hayles, N. Katherine (2007), 'Hyper and Deep Attention: The Generational Divide in Cognitive Modes', *Profession*: 187–99.

Heidegger, Martin (1962) [1927], *Being and Time*, trans. John Macquarrie and Edward Robinson, Oxford: Blackwell.

—(1993) [1949], 'The Question Concerning Technology', in *Basic Writings*, ed. David Farrell Krell, London: Routledge.

—(1997) [1929], *Kant and the Problem of Metaphysics*, trans. Richard Taft, Bloomington: Indiana University Press.

—(1998) [1929], 'What is Metaphysics?', in *Pathmarks*, ed. W. McNeill, trans. D. F. Krell, Cambridge: Cambridge University Press.

Hill, Leslie (1997), *Blanchot: Extreme Contemporary*, London: Routledge.

Houellebecq, Michel (2001) [1998], *Atomised*, trans. Frank Wynne, London: Vintage; *Les Particules élémentaires*, Paris: Flammarion, 2008.

—(2006), *The Possibility of an Island*, trans. Gavin Bowd, London: Phoenix; *La Possibilité d'une île*, Paris: Fayard, 2005.

Howells, Christina (2011), *Mortal Subjects: Passions of the Soul in Late Twentieth-Century French Thought*, Cambridge: Polity.

Husserl, Edmund (1991) [1928], *On the Phenomenology of the Consciousness of Internal Time (1893–1917)*, trans. John Barnett Brough, London: Kluwer Academic Press.

Jacob, François (1974), *The Logic of Life: A History of Heredity*, trans. Betty E. Spillmann, London: Allen Lane; *Logique du vivant: Une Histoire de l'hérédité*, Paris: Gallimard, 1970.

James, Ian (2010), 'Bernard Stiegler and the Time of Technics', *Cultural Politics*, 6.2: 207–28.

—(2012), *The New French Philosophy*, Cambridge: Polity.

Johnson, Christopher (2011), 'Leroi-Gourhan and the limits of the human', *French Studies*, 65.4: 471–87.

Kant, Immanuel (2003) [1781/1787], *Critique of Pure Reason*, trans. Norman Kemp Smith, New York: Palgrave Macmillan.

—(2007) [1790], *Critique of Judgement*, trans. James Creed Meredith, revised and edited by Nicholas Walker, Oxford: Oxford University Press; *Kritik der Urteilskraft*, Akademie-Ausgabe, Volume V.

Knight, Chris, Michael Studdert-Kennedy and James R. Hurford, eds (2000), *The Evolutionary Emergence of Language. Social Function and the Origins of Linguistic Form*, Cambridge: Cambridge University Press.

Lacan, Jacques (1978), *The Four Fundamental Concepts of Psycho-analysis: The Seminar of Jacques Lacan, Book XI: 1963–4*, ed. Jacques-Alain Miller, trans. Alan Sheridan, New York: Norton.

—(2006a) [1966], *Écrits: The First Complete Edition in English*, trans. Bruce Fink, New York: W. W. Norton; *Écrits*, Paris: Seuil.

—(2006b) [1971], *Le Séminaire de Jacques Lacan, livre XVIII. D'un discours qui ne serait pas du semblant*, Paris: Seuil.

Laplanche, Jean and Jean-Bertrand Pontalis (1973), *The Language of Psychoanalysis*, trans. Donald Nicholson-Smith, New York: Norton; *Vocabulaire de la Psychanalyse*, Paris: PUF, 1967.

Lawlor, Leonard (2003), 'The Beginnings of Thought: The Fundamental Experience in Derrida and Deleuze', in *Between Deleuze and Derrida*, eds Paul Patton and John Protevi, London: Continuum, pp. 67–83.

LeDoux, Joseph (2003), *Synaptic Self: How our Brains Become Who We Are*, Harmondsworth: Penguin.

Leroi-Gourhan, André (1964), *Les Religions de la préhistoire*, Paris: PUF.

—(1965), *Préhistoire de l'art occidental*, Paris: Mazenod.

—(1971) [1943], *Évolution et techniques, 1: L'Homme et la matière*, Paris: Albin Michel.

—(1973) [1945], *Évolution et techniques, 2: Milieu et techniques*, Paris: Albin Michel.

—(1982) *Les Racines du monde. Entretiens avec Claude-Henri Rocquet*, Paris: Belfond.

—(1989) [1983], *The Hunters of Prehistory*, trans. Claire Jacobson, New York: Atheneum.

—(1993) *Gesture and Speech*, trans. Anna Bostock Beger, Boston: The

MIT Press; *Le Geste et la parole, 1. Technique et langage*, Paris: Albin Michel, 1964; *Le Geste et la parole, 2. La Mémoire et les rythmes*, Paris: Albin Michel, 1965.

Lévi-Strauss, Claude (1978), *Introduction to the Work of Marcel Mauss*, trans. Felicity Baker, London: Routledge; 'Introduction à l'œuvre de Marcel Mauss', in *Sociologie et anthropologie*, Paris: PUF, 1950.

—(1992), *Tristes tropiques*, trans. John and Doreen Weightman, New York: Penguin; *Tristes tropiques*, Paris: Plon, 1955.

—(2009), 'La Biologie, science exemplaire', in *Levi-Strauss par Lévi Strauss*, special number of *Le Nouvel Observateur*, 74 (November–December): 48–51.

Lindon, Mathieu (2011), *Ce qu'aimer veut dire*, Paris: POL.

Loewald, Hans (1988), *Sublimation: Inquiries into Theoretical Psychoanalysis*, New Haven: Yale University Press.

Luncz, Lydia et al. (2012), 'Evidence for Cultural Differences between Neighbouring Chimpanzee Communities', *Current Biology*, 22.10: 1–5.

Lyotard, Jean-François (1984), *The Postmodern Condition: A Report on Knowledge*, trans. Geoff Bennington, Minneapolis: Minnesota University Press; *La Condition postmoderne*, Paris: Minuit, 1979.

—(1991) [1988], *The Inhuman: Reflections on Time*, trans. Geoffrey Bennington and Rachel Bowlby, Cambridge: Polity.

—(2004), *Libidinal Economy*, trans. Iain Hamilton Grant, London: Continuum; *Économie libidinale*, Paris: Minuit, 1974.

Malabou, Catherine (2005), *The Future of Hegel: Plasticity, Temporality and Dialectic*, trans. Lisabeth During, London: Routledge; *L'Avenir de Hegel: plasticité, temporalité, dialectique*, Paris: Vrin, 1996.

—(2008), *What Should we do with our Brain?*, trans. Sebastian Rand, New York: Fordham University Press; *Que faire de notre cerveau?*, Paris: Bayard, 2004.

—(2011), *The Heidegger Change: On the Fantastic in Philosophy*, ed. and trans. Peter Skafish, New York: SUNY Press; *Le Change Heidegger: du fantastique en philosophie*, Paris: Léo Scheer, 2004.

—(2012), *The New Wounded*, trans. Steven Miller, New York, Fordham University Press; *Les Nouveaux blessés: de Freud à la neurologie, penser les traumatismes contemporains*, Paris: Bayard, 2007.

Marcuse, Herbert (1964), *One-Dimensional Man: Studies in the Ideology of Advanced Industrial Society*, London: Routledge.

—(1966) [1955], *Eros and Civilization: A Philosophical Inquiry into Freud*, London: Allen Lane.

Marx, Karl (1976) [1867], *Capital, Volume 1*, London: Penguin.

Miller, James (1993), *The Passion of Michel Foucault*, London: Harper Collins.

Moore, Gerald (2011), 'Gay Science and (No) Laughing Matter: The Eternal Returns of Michel Houellebecq', *French Studies*, 75.1: 45–60.

—(2012), 'Crises of Derrida: Theodicy, Sacrifice and (Post-) Deconstruction', *Derrida Today*, 5.2: 264–82.

—(2013), 'Embers of the Sublime: Sacrifice and the Sensation of Existence', *Senses and Society*, 8.1: 37–49.

Nagle, D. Brendan (2005), *Household and City in the Writings of Aristotle*, Cambridge: Cambridge University Press.

Nealon, Jeffrey (2007), *Foucault Beyond Foucault: Power and its Intensifications since 1984*, Stanford: Stanford University Press.

Nietzsche, Friedrich (1969) [1885], *Thus Spoke Zarathustra*, trans. R. J. Hollingdale, Harmondsworth: Penguin.

Noys, Benjamin (2010), 'Apocalypse, Tendency, Crisis', *Mute: Culture and Politics after the Net*, 2.15, <www.metamute.org/editorial/articles/apocalypse-tendency-crisis> (last accessed 21 March 2013).

Pinker, Steven (2002), *The Blank Slate. The Modern Denial of Human Nature*, London: Penguin.

Preciado, Beatriz (2008), *Testo Junkie: sexe, drogue et biopolitique*, Paris: Grasset.

Rancière, Jacques (1991) [1987], *The Ignorant Schoolmaster*, trans. Kristin Ross, Stanford: Stanford University Press.

—(2003) [1983], *The Philosopher and His Poor*, trans. John Drury, Corinne Oster and Andrew Parker, ed. Andrew Parker, Durham, NC: Duke University Press.

Roberts, Ben (2005), 'Stiegler Reading Derrida: The Prosthesis of Deconstruction in Technics', *Postmodern Culture*, 16:1, <http://muse.jhu.edu/journals/postmodern_culture/toc/pmc16.1.html> (last accessed 21 March 2013).

—(2006), 'Cinema as mnemotechnics: Bernard Stiegler and the industrialisation of memory', *Angelaki*, 11.1: 55–63.

—(2012), 'Technics, Individuation and Tertiary Memory: Bernard Stiegler's challenge to media theory', *New Formations*, 77: 8–20.

Rodowick, David Norman (1994), *The Crisis of Political Modernism*, Berkeley: University of California Press.

Rose, Jacqueline (2002) [1996], 'Of Knowledge and Mothers', in *The Vitality of Objects: Exploring the Work of Christopher Bollas*, ed. Joseph Scalia, Middletown: Wesleyan University Press, 108–25.

Rose, Steven (2005), *Lifelines: Life Beyond the Gene*, expanded edition, London: Vintage.

Ross, Daniel (2007), 'Politics, Terror, and Traumatypical Imagery', in *Trauma, History, Philosophy*, Matthew Sharpe, Murray Noonan and Jason Freddi, eds, Cambridge: Cambridge Scholars Publishing.

Rousseau, Jean-Jacques (1984) [1755], *A Discourse on Inequality*, trans. Maurice Cranston, Harmondsworth: Penguin.

Sahlins, Marshall (1976), *The Use and Abuse of Sociobiology: An Anthropological Critique of Sociobiology*, London: Tavistock.

Saper, Craig (2009), 'Sublimation as Media: inter urinas et faeces nascimur', *Discourse*, 31.1–2:51–71.

Sartre, Jean-Paul (1960), *Critique de la raison dialectique*, Paris: Gallimard.

— (1971), *L'Idiot de la famille, I*, Paris: Gallimard.

Schlanger, Nathan (2004), ' "Suivre les gestes, éclat par éclat" – la chaîne opératoire d'André Leroi-Gourhan', in *Autour de l'homme. Contexte et actualité d'André Leroi-Gourhan*, Françoise Audouze and Nathan Schlanger, eds, Antibes: APDCA, pp. 127–47.

Simondon, Gilbert (1989) [1958], *Du mode de l'existence des objets techniques*, Paris: Aubier, <http://english.duke.edu/uploads/assets/Simondon_MEOT_part_1.pdf> (last accessed 21 March 2013).

—(1992) [1964], 'The Genesis of the Individual', in *Incorporations*, Jonathan Crary and Sanford Kwinter, eds, New York: Zone, 1992.

—(2007) [1958], *L'Individuation psychique et collective: à la lumière des notions de forme, information, potentiel et métastabilité*, Paris: Aubier.

Sinnerbrink, R. (2009), 'Culture Industry Redux: Stiegler and Derrida on Technics and Cultural Politics', *Transformations*, 17, <www.transformationsjournal.org/journal/issue_17/article_05.shtml> (last accessed 21 March 2013).

Spencer, Herbert (1855), *The Principles of Psychology*, London: Longman and Brown.

Stiegler, Barbara (2001), *Nietzsche et la biologie*, Paris: Presses Universitaires de France.

Stiegler, Bernard (1996), 'Persephone, Oedipus, Epimetheus', trans. Richard Beardsworth, *Tekhnema: Journal of Philosophy and Technology*, 3: 69–112.

—(2007a), 'Technoscience and Reproduction', Parallax 13.4: 29–45.

—(2007b), 'L'Inquiétante étrangeté de la pensée et la métaphysique de Pénélope', preface to Gilbert Simondon, *L'Individuation psychique et collective: à la lumière des notions de forme, information, potentiel et métastabilité*, Paris: Aubier.

—(2007c), 'Questions de pharmacologie générale. Il n'y a pas de simple pharmakon', *Psychotropes*, 13.3–4: 27–54, <www.cairn.info/revue-psychotropes-2007-3-page-27.htm> (last accessed 21 March 2013).

—(2008a), 'Take Care (Prendre Soin)', presentation at *To Take Care*, colloquium 'On Nature as Culture', <www.arsindustrialis.org/node/2925> (last accessed 21 March 2013).

—(2008b), 'L'Être soigneux', in *L'Avenir du passé*, eds Bernard Stiegler and Jean-Paul Demoule, Paris: Cairn, pp. 13–25.

—(2009), 'Teleologics of the Snail: The Errant Self Wired to a WiMax Network', *Theory, Culture and Society*, 26.2–3: 33–45.

—(2011a), 'Faire du cinéma' (unpublished paper), *Impact of Technology on the History and Theory of Cinema*, Université de Montreal/McGill University, Montreal, Nov. 2011.

—(2011b), '"This System Does Not Produce Pleasure Anymore": An Interview with Bernard Stiegler', interviewer Peter Lemmons, *Krisis: Journal for Contemporary Philosophy*, 1: 33–41.

—(2012a), 'Relational Ecology and the Digital Pharmakon', trans. Patrick Crogan. *Culture Machine*, 13: 1–19.

—(2012b), 'An Organology of Dreams' (unpublished paper), Film and Philosophy Conference, Queen Mary University/Kings College London, London, September 2012.

Turner, Chris (2010), 'Kant avec Ferry: Some Thoughts on Bernard Stiegler's *Prendre soin, I. De la jeunesse et des générations* (Paris: Flammarion, 2008)', *Cultural Politics*, 6.2: 253–8.

Vernant, Jean-Pierre (1989), 'At Man's Table: Hesiod's Foundation Myth of Sacrifice', in *The Cuisine of Sacrifice among the Greeks*, eds Marcel Detienne and Jean-Pierre Vernant, trans. Paula Wissing, Chicago: Chicago University Press; 'À la table des hommes', in *La Cuisine du sacrifice en pays grec*, Paris: Gallimard, 1979.

de Waal, Frans (2009), *The Age of Empathy: Nature's Lessons for a Kinder Society*, London: Souvenir.

Weber, Max (1958) [1905], *The Protestant Ethic and the Spirit of Capitalism*, New York: Scribner.

Weismann, August (2008) [1904], 'The Selection Theory', in *Evolution in Modern Thought*, Haeckel, Thomson, Weismann et al., New York: Forgotten Books/The Modern Library.

Wills, David (2006), 'Techneology and the Discourse of Speed', in *The Prosthetic Impulse: From a Posthuman Present to a Biocultural Future*, eds Marquand Smith and Joanne Morra, Cambridge, MA: MIT Press.

Winnicott, D. W. (2005) [1971], *Playing and Reality*, London: Routledge.

—(2011) [1941] 'The Observation of Infants in a Set Situation', in

Reading Winnicott, ed. Lesley Caldwell and Angela Joyce, London: Routledge, pp. 33–53.

Wray, Alison, ed. (2002), *The Transition to Language*, Oxford: Oxford University Press.

Žižek, Slavoj (1994), *The Metastases of Enjoyment: Six Essays on Women and Causality*, London: Verso.

—(2008), *In Defence of Lost Causes*, London: Verso.

Index